PHASE-LOCKED LOOP FREQUENCY SYNTHESIS METHODS

For a complete listing of titles in the
Artech House Microwave Library,
turn to the back of this book.

PHASE-LOCKED LOOP FREQUENCY SYNTHESIS METHODS

Vitaly Koslov

**ARTECH
HOUSE**

BOSTON | LONDON
artechhouse.com

Library of Congress Cataloging-in-Publication Data
A catalog record for this book is available from the U.S. Library of Congress.

British Library Cataloguing in Publication Data
A catalogue record for this book is available from the British Library.

Cover design by Iain Hill

ISBN 13: 978-1-68569-101-1

© **2026**
Artech House
685 Canton Street
Norwood, MA 02062

10 9 8 7 6 5 4 3 2 1

CONTENTS

3

SCHEMES WITH THE FRACTIONAL DIVIDER AND WITH THE SUPPRESSION OF FRACTIONAL NOISE AND OTHER SCHEMES   43

4

THE IDEA OF A MULTIFREQUENCY PHASE DETECTOR   79

5

SYNTHESIZERS OF PHASE DIGITAL SYNTHESIZERS, AND PHASE DIGITAL SYNTHESIZERS WITH DELTA SIGMA MODULATION TYPES   91

6
CONCLUSION   131

PREFACE

A frequency synthesizer is a necessary block of modern telecommunication and measuring systems and largely determines their main characteristics. To ensure high-quality communication and measurements, the synthesizer itself must meet high requirements, the most important of which are the spectral purity of the generated signal, frequency switching speed, and frequency resolution (frequency step size). Also important, especially for mobile systems, are low power consumption, small size and weight, and low cost. Therefore, the main task for the developers of such systems is to find ways to meet these requirements as much as possible.

There are many fundamental publications, for example [1–17], which consider both the theoretical foundations of frequency synthesis and the issues of practical construction of such systems. Without diminishing the merits of the authors of these works and without questioning the importance of the material presented by them and their undoubted usefulness for the developers of equipment of this class, it should, nevertheless, be recognized that, in these works, there is no systematic analysis of the improvement of frequency synthesis systems over time.

This is a path from the simplest structure of a single-loop synthesizer with a frequency divider with an integer variable division ratio, through the transformation of the latter into a frequency divider with a variable fractional division ratio and for fractional noise, through complicated multiloop circuits and again to single-loop circuits, but already at a higher level with phase splitting and using delta-sigma modulation.

In this monograph, an attempt is made to fill this gap, with the main attention paid to synthesis systems based on phase-locked loop (PLL) as the most promising direction, which has received the widest application in synthesizers of telecommunications and measuring equipment.

This material may turn out to be interesting and useful both for the developers of radio equipment and for students of the relevant specialties.

References

[1]   Manassewitch, V., *Frequency Synthesizers: Theory and Design*, Third Edition, New York: Wiley, 2005.

[2]   Kroupa, V., *Frequency Synthesis Theory, Design and Applications*, New York: John Wiley & Sons, 1973.

[3]   Rohde, U. L., *Digital PLL Frequency Synthesizers: Theory and Design*, Upper Saddle River, NJ: Prentice Hall, 1983.

[4]   Rohde, U. L., *Microwave and Wireless Synthesizers: Theory and Design*, New York: John Wiley & Sons, 1997.

[5]   Best, R. E., *Phase-Locked Loops: Theory, Design, and Applications*, New York: McGraw-Hill, 1984.

[6]   Goldberg, B. -G., *Digital Frequency Synthesis Demystified*, Eagle Rock, VA: LLH Publishing, 1999.

[7]   Egan, W., *Frequency Synthesis by Phase Lock*, Second Edition, New York: Wiley, 1999.

[8]   Egan, W. F., *Phase–Lock Basics*, Second Edition, New York: John Wiley & Sons, 2007.

[9]   Gardner, F., *Phaselock Techniques*, Third Edition, New York: Wiley, 2005.

[10] Rohde, U. L., and D. P. Newkirk, *RF/Microwave Circuit Design for Wireless Applications*, New York: John Wiley & Sons, 2000.

[11] Rogers, J., C. Plett, and F. Dai, *Integrated Circuit Design for High-Speed Frequency Synthesis*, Norwood, MA: Artech House, 2007.

[12] Crawford, J., *Advanced Phase-Lock Techniques*, Norwood, MA: Artech House, 2008.

[13] Chenakin, A., *Frequency Synthesizers: Concept to Product*, Norwood, MA: Artech House, 2010.

[14] Chenakin, A. V., and A. V. Gorevoy, *Practical Construction of Microwave Frequency Synthesizers*, Moscow: Hotline-Telecom, 2021, in Russian.

[15] Zaretsky, M. M., and M. E. Movshovich, *Frequency Synthesizers with a Phase-Locked Loop*, Moscow: Energiya, 1974, in Russian.

[16] Shapiro, D. N., and A. A. Panin, *Fundamentals of the Theory of Frequency Synthesis*, Moscow: Radio and Communications, 1981, in Russian.

[17] Ryzhkov, A. V., and V. N. Popov, *Frequency Synthesizers in Radio Communication Technology*, Moscow: Radio and Communications, 1991, in Russian.

INTRODUCTION

One of the most important tasks in the construction of a frequency synthesizer is the simultaneous provision of sufficiently high characteristics of both the spectral purity of the signal and speed of frequency switching. To solve this task, resort to complex structures that are expensive and have significant energy consumption. However, there has always been a desire to make it as simple as possible, for example, in a single-loop phase-locked loop (PLL) or in a direct digital synthesizer (DDS).

This work considers both early not very successful attempts in this direction, as well as later, more successful technical solutions. Consideration begins with the simplest, one-loop structure with a frequency divider with a divider of integer variable ratio (DIVR) [1, 2], which has significant drawbacks, due to which the area of its possible application as an independent device is very limited. Nevertheless, there are interesting ideas on how to get relatively good results using a few simple PLLs based on the variable division ratio.

A powerful incentive to improve the characteristics of single-loop frequency synthesizers was made by Loposer, who proposed using a frequency divider with a fractional variable division factor in the PLL [3], followed by a number of works analyzing the possibilities of such a structure, for example, [4, 5]. This made it possible to significantly increase the comparison frequency in the synthesizer, while maintaining a high-frequency resolution, to expand the loop bandwidth, that is, to increase the frequency switching speed. At the same time, technical solutions were required to suppression of the fractional interference created by the fractional divider. Examples of such solutions will be given in the relevant sections.

The essence of suppression lies in the formation of a correcting signal of the same shape and magnitude as the fractional interference in the control circuit of a voltage-controlled oscillator (VCO), but of the opposite phase. Together with the interference, this correcting signal gives only a constant component, and, thus, the fractional interference is eliminated. However, it is sometimes difficult and often impossible to maintain the amplitude and shape of the suppressing signal with the required accuracy in a wide frequency range of the synthesizer and to obtain a sufficiently accurate analog summing. Therefore, in order to achieve an acceptable, small, residual level of fractional noise, it is necessary, again, to build multiloop systems.

It is important to note here that the aforementioned basic characteristics of a one-loop synthesizer depend on the method of phase comparison of the reference and tuned signals. Is it required to bring their frequencies to equality, and, if so, how is this achieved, or are there other methods of phase detection directly at unequal frequencies? When considering PLL-based synthesizers, special attention will be paid to the possible options for constructing the phase detector (PD).

The following analysis of various innovations in the field of frequency synthesis ends with a consideration of a new structure of a single-loop synthesizer, the implementation of which in an integrated chip will allow achieving extremely high characteristics of spectral purity and agility.

The material is presented in the most accessible form, without excessive mathematization, clearly illustrated with diagrams to explain the work of the circuits, which can improve its perception by a wide range of readers.

References

[1]   Young, C. J., "Stabilized Oscillator Generator," US Patent 2,490,500, December 6, 1949, filed December 29, 1946.

[2]   Woodward, J. D., "Variable Frequency Oscillation Generator," US Patent 2,490,499, December 6, 1949, filed April 23, 1947.

[3] Loposer, T. L., "Frequency Synthesizer Using Fractional Division by Digital Techniques within a Phase-Locked Loop," US Patent 3,353,104, November 14, 1967, filed October 2, 1963.

[4] Varfolomeev, G. F., "The Spectrum of Fractional Interference in a PLL with a Fractional Frequency Divider," *Communication Equipment,* Ser. TRS, Issue 10(21), 1978, in Russian.

[5] Romanov, S. K., "Determination of Interference in the PLL System with a Fractional Frequency Divider in the Feedback Circuit," *Theory and Technology of Radio Communication: Scientific and Technical Collection/VNIIS,* Issue 2, Voronezh, 2003, in Russian.

1

SCHEMES WITH DIVIDERS OF INTEGER
VARIABLE DIVISION RATIOS

1.1 THE SIMPLEST ONE-LOOP STRUCTURE

For the first time, a single-loop frequency synthesizer with the divider of integer variable ratio (DIVR) was patented, almost simultaneously, by the Americans Young [1] and Woodward [2]. True, the schemes given in the descriptions of the inventions are rather peculiar, they are depicted on the elements of the then technological level that are not used now, but nevertheless the idea of such a technical solution is quite clear and can be illustrated by a simplified scheme presented in Figure 1.1. According to the adopted later terminology, such a structure is now referred to as an integer-N phase-locked loop (PLL) synthesizer.

The synthesizer contains a voltage-controlled oscillator (VCO), operating in the required signal frequency range. It is covered by negative feedback through the divider with a controlled division ratio N and also contains a phase detector (PD) and a lowpass filter. After the lowpass filter, a direct current amplifier (DCA) can also be installed to obtain the required control voltage swing. A reference signal is supplied to the other input of the PD, the frequency of which F_{PD} is equal

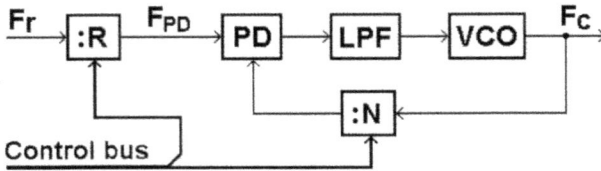

Figure 1.1   Integer-N PLL synthesizer.

to the required frequency step size. A signal error is generated in the PD, which, through a lowpass filter, enters the VCO control circuit, bringing its frequency F_c to the equality

$$F_c = NF_{PD}$$

The frequency F_{PD} used for comparison in a PD can be obtained by dividing the frequency F_r of the reference source by R times. Then the expression for the frequency F_c at the synthesizer output can be written in the form

$$NF_r/R$$

As an example, let's assume that it is required to obtain a frequency with a step size 10 kHz in the frequency range F_c = 700 to 800 MHz using the reference frequency F_r equal to F_r = 10 MHz. Then the division factor R should be chosen equal to R = 1,000 so that the comparison frequency F_{PD} would be equal to 10 kHz. It is clear that assigned task will be solved by choosing the N coefficient in the range from 70,000 to 80,000.

An obvious advantage of the considered scheme is its exceptional simplicity. However, there are also very significant disadvantages. Interference from the PD output modulates the VCO, creating discrete interference side bands in the signal spectrum. To suppress them, the bandwidth of the lowpass filter should be at least an order of magnitude less than the comparison frequency F_{PD}. This significantly limits the speed of frequency switching and capabilities in suppression of VCO noise.

In addition, there is another problem with the spectral purity of the signal. The expression for the phase noise at the synthesizer output within the PLL bandwidth can be written as

$$G = G_{PD} + 20 \lg N$$

where the G_{PD} is the noise of the PD itself and the noise of the reference source and the frequency dividers by N and R are recalculated to the PD input. Accordingly, the noise spectrum of the signal deteriorates significantly if one wants to get a fine frequency step size dF, increasing the coefficient N. Also, due to the corresponding narrowing of the PLL bandwidth, the noise of the VCO is weakly suppressed, which makes an additional contribution to the degradation of the signal spectrum.

This problem can be somewhat mitigated by making both division ratios R and N controllable. This makes it possible to obtain a frequency with a finer frequency resolution. This can be shown using Table 1.1, which summarizes the values of R, F_{PD}, N, dF, and the resulting frequency F_c.

As can be seen from Table 1.1, the step size dF decreases by three orders of magnitude, but the range of possible values of the F_c frequency is also reduced to such an extremely small value as only

Table 1.1

R	F_{PD}, Hz	N	F_c, MHz	dF, Hz
1000	10000.000000	70001	700.01000	10
1001	9990.009990	70071	700.009990	10
1002	9980.039920	70141	700.008980	10
1143	8748.906387	80011	700.008749	8
1144	8741.258741	80081	700.008741	8
1999	5002.501251	139931	700.005003	2.5
2000	5000.000000	14001	700.005000	2.5

5 kHz. However, such unique cases are not excluded in which the described idea can find its embodiment.

At the same time, the frequency range F_c can be obtained even wider if the values of the coefficient R are not chosen as large as shown in Table 1.2. There, this coefficient is reduced by an order of magnitude, due to which the frequency range F_c is also expanded by an order of magnitude, up to 50 kHz. However, at the same time, the step size became much larger in comparison with the previous case, its maximum value became equal to 990 Hz, and yet it turned out to be an order of magnitude smaller than in the variant with a constant value of $F_{PD} = 10$ kHz. In addition, the comparison frequency in the PD has increased by an order of magnitude and, accordingly, the multiplication factor of the noise reduced contributing to the improvement of both the spectral purity and the switching speed of the synthesizer.

From the tables given for the cases of a controlled coefficient N, it can be seen how is simple the algorithm for choosing this coefficient. So that there are no gaps when tuning in the frequency range, with each change in R by 1, the coefficient N changes by 70 units. It can be also noticed that, if selecting only the upper frequencies of the range, then the step size there is significantly reduced.

In favor of any of the variants of the one-loop synthesizer, there is the fact that, in the noise spectrum at the PD output, there are only spurs with the reference frequency and their harmonics, and there are no combinations of other frequencies accompanying the constant component. This makes filtering spurs easier.

Some improvement in frequency resolution can be achieved by using fractional R and N ratios with filtering the fractional components in the PLL. However, at the same time, the bit depth of the

Table 1.2

R	F_{PD}, Hz	N	F_C, MHz	dF, Hz
100	100000.000000	7001	700.100000	990
101	99009.900990	7071	700.099010	990
102	98039.215686	7141	700.098039	971
103	97087.378641	7211	700.097087	952
199	50251.256281	13931	700.050251	251
200	50000.000000	14001	700.050000	251

fractions should not be high, so that narrowing the PLL bandwidth for filtering them does not lead to a significant decrease in the synthesizer's performance.

1.2   SCHEME WITH A FREQUENCY MIXER

To improve the spectral characteristics of a single-loop synthesizer, a frequency mixer, MX, can be used, which is included in the feedback loop, as shown in Figure 1.2. A reference frequency F_r multiplied by M times is fed to one of the mixer inputs. The signal of the difference frequency $F_c - MF_r$ is filtered by a bandpass filter (BPF) and enters the DIVR input with a division ratio N.

In accordance with the presented scheme, the frequency formation looks like

$$F_c = (N/R + M)F_r$$

If it is assumed that $F_r = 10$ MHz, $R = 1{,}000$, $M = 60$, and $N = 10{,}000$ to $19{,}999$, then the above formula gives the result $F_c = 700$ to 799.99 MHz, and the step size remains the same, that is, $dF = 10$ kHz. However, in this scheme, the division ratio in the PLL has decreased by an order of magnitude, which means that the gain of the noise is reduced to the input of the PD by the same factor. This is the main advantage of the scheme. It is also important that the DIVR in this

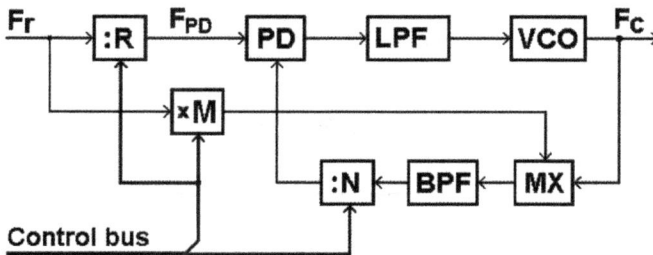

Figure 1.2   Scheme with a frequency mixer in a PLL.

scheme operates at a reduced frequency, which is why it is cheaper and less than the power consumption.

The disadvantage of this technical solution lies in the comparative complexity of the scheme and the possibility of the formation of additional interference of a combinational nature at its output generated in the mixer and passing further along the loop into the frequency control circuit of the VCO. Therefore, a careful design of the mixer and frequency multiplier assemblies is required.

1.3   TOLLEFSON'S SCHEME

Noteworthy is the structure proposed by Tollefson [3] and shown in Figure 1.3. These are two PLLs: PLL-1 and PLL-2, connected to each other through the mixer, MX. Each of them contains a VCO, DIVR, PD, and a lowpass filter, indicated by the numbers of the corresponding

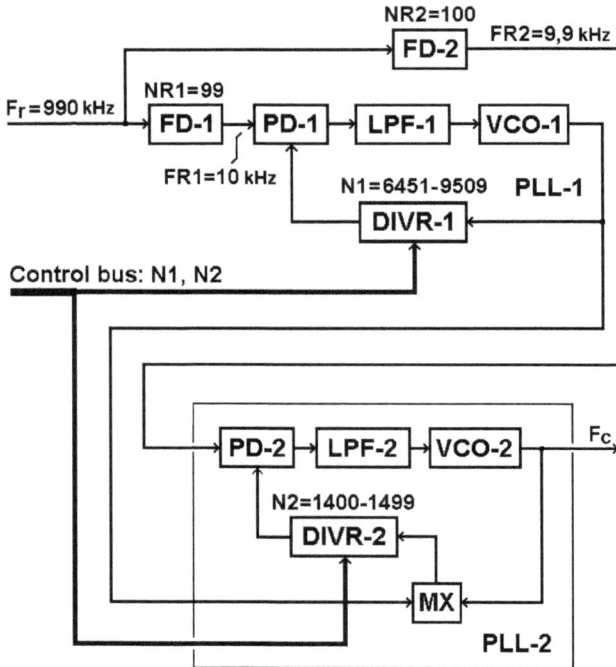

Figure 1.3 Tollefson's scheme.

PLL. The comparison frequencies FR1 and FR2 are different for them, but they are obtained from the same reference source of frequency F_r using the corresponding frequency dividers FD-1 and FD-2 with division coefficients NR1 and NR2, respectively.

To make it easier to understand how the scheme works, a numerical example of its parameters is given on it. At the frequency of the reference oscillator equal to $F_r = 990$ kHz, the comparison frequencies FR1 and FR2 in the corresponding PLL, obtained using frequency dividers FD-1 and FD-2 with the corresponding ratios NR1 and NR2, are equal to FR1 = 990/99 = 10 kHz and FR2 = 990/100 = 9.9 kHz. In this case, the output of the synthesizer provides a frequency step size equal to 100 Hz, which is equal to the difference between the comparison frequencies FR1 and FR2. Let's show this with numerical examples.

Based on the given structure of the synthesizer, the frequency at its output can be calculated by the formula

$$F_c = FR1 \times N_1 - FR2 \times N_2$$

Then the lower frequency of the synthesizer range, obtained with the coefficients $N_1 = 6,451$ and $N_2 = 1,400$, is equal to 50.650 MHz. To obtain the next frequency in the range, it is necessary to simultaneously shift the ratios N_1 and N_2 by one (this is the algorithm for controlling these coefficients), and, as a result, the signal frequency will be equal to 50.6501 MHz, which is 100 Hz higher than the previous one. Then, with each addition of units in the coefficients N_1 and N_2, according to the specified algorithm, 10 kHz is added to the signal frequency due to the action of the first loop and 9.9 kHz is subtracted due to the action of the second loop, and as a result, the signal frequency increases with a step of 100 Hz.

Using the above formula, it is easy to calculate the upper frequency of the range. It is obtained with the coefficients $N_1 = 9,509$ and $N_2 = 1,499$ and is equal to 80.2499 MHz.

The advantage of the scheme is the ability to obtain a small step size of the frequency at relatively high comparison frequencies. However, these possibilities are limited by the difficulty of obtaining two frequencies from the reference source, when these frequencies are high enough and the difference between them is small. For example, if there is need to have a comparison frequency of about 1 MHz and a step size of 1 Hz, you need to get two frequencies of 1 MHz and 1.000001 MHz, which is very difficult.

It should also be noted that Tollefson's scheme actually follows from the method proposed by Denisov 11 years earlier [4].

1.4 MARTIN'S SCHEME

Also interesting is Martin's idea [5], schematically shown in Figure 1.4. The voltage controlled generator, VCO-1, is included into a wideband PLL with DIVR-1, which has relatively low division ratios. The loop also includes the PD, PD-1, and the lowpass filter, LPF-1. The reference frequency for this loop is taken from VCO-2 included into a second narrowband PLL with DIVR-2 of relatively large division ratios and a rather low comparison frequency. The second loop also includes the PD, PD-2, and the lowpass filter, LPF-2. However, it is impossible to call the second loop separate or independent since DIVR-2

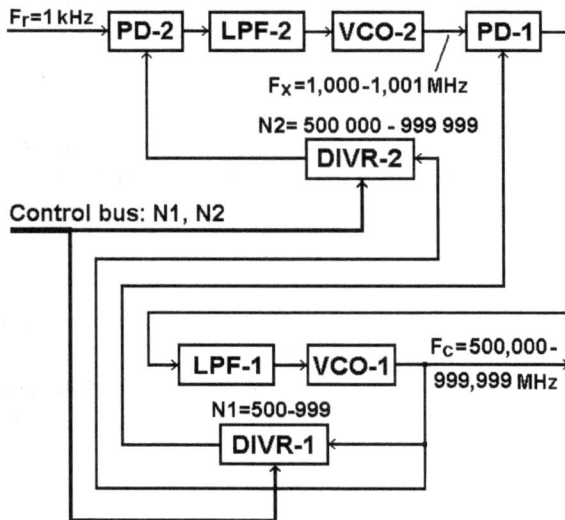

Figure 1.4 Martin's scheme.

operates from the same VCO-1 and the second loop works through the first loop. Therefore, some developers call such a structure a "one and a half loop" and sometimes "tandem." Both DIVRs are conjugated so that the most significant digits in them are switched simultaneously. Figure 1.4 also shows an example of obtaining an octave range in such a scheme.

The signal frequency at the output of the synthesizer is $F_c = F_r \times N_2 = 500{,}000$ to $999{,}999$ MHz and is switchable in 1-kHz steps. The reference frequency for the first loop is in the range $F_x = F_r \times N_2/N_1 = 1.000$ to 1.001 MHz.

In such a structure, the tuning range of VCO-2 is much smaller than that in VCO-1, and therefore its high spectral purity can be ensured even in a narrowband PLL. Since the frequency of its signal is the reference for the wideband loop operating at the output, an improvement in the signal spectrum at the output of the synthesizer is also achieved. The advantage of the proposed scheme also lies in the possibility of fast frequency switching in large steps. The problem of the low agility of the system when switching the frequency in small steps remains unresolved.

1.5 CASCADE PLL SYNTHESIZER

The principle of such a frequency synthesis can be explained using the scheme shown in Figure 1.5. It shows two PLLs connected in series. Both loops are identical and each contains a VCO, a DIVR, a PD, and a lowpass filter. Each of the named blocks is numbered on the diagram in accordance with the loop number. PLL-1 and PLL-2 are connected to each other through DIVR-3. The first of them, with the participation of DIVR-3, forms a variable reference frequency for the second one, which completes the structure of the synthesizer. To clarify the operation of such a structure, a numerical example is shown in Figure 1.5.

Next are some calculations for the frequency F_c at the synthesizer output with the reference frequency $F_r = 50$ MHz:

$N_1 = 60 \quad F_1 = 3{,}000 \quad N_3 = 59 \quad N_2 = 30 \quad F_c = 1{,}525{,}424 \text{ MHz}$

$N_1 = 59 \quad F_1 = 2{,}950 \quad N_3 = 58 \quad N_2 = 30 \quad F_c = 1{,}525{,}862 \text{ MHz}$

$N_1 = 58 \quad F_1 = 2{,}900 \quad N_3 = 57 \quad N_2 = 30 \quad F_c = 1{,}526{,}316 \text{ MHz}$

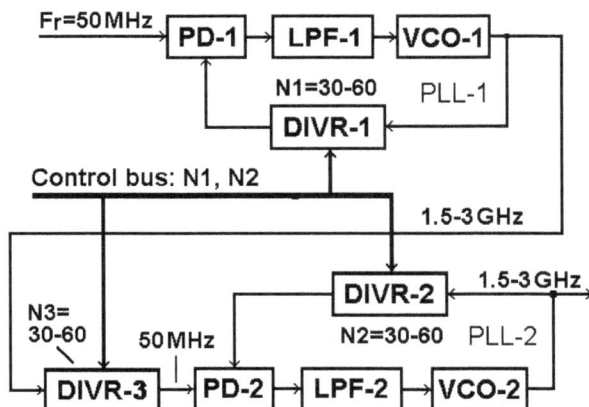

Figure 1.5 Sequential connection of two PLLs.

and

$$N_1 = 59 \quad F_1 = 2{,}950 \quad N_3 = 58 \quad N_2 = 58 \quad F_c = 2{,}950{,}000 \text{ MHz}$$
$$N_1 = 58 \quad F_1 = 2{,}900 \quad N_3 = 57 \quad N_2 = 58 \quad F_c = 2{,}950{,}877 \text{ MHz}$$
$$N_1 = 57 \quad F_1 = 2{,}850 \quad N_3 = 56 \quad N_2 = 58 \quad F_c = 2{,}951{,}786 \text{ MHz}$$

As can be seen from the above calculations, the resulting frequency step size turns out to be much less than the initial reference frequency F_r. In the lower part of the synthesized frequency range, it is about 450 kHz, and in the upper part, it is about 900 kHz.

It is advisable to choose the division factors N_1 and N_3 with values close to each other (in the example considered, they differ by one). Then the generated reference frequency F_2 for the second loop differs little from the original reference frequency F_r, that is, it also turns out to be quite high. Due to this, the bandwidth of the second loop can be chosen as wide as that of the first loop.

It can also be noted that the frequency range formed by the first loop does not have to be equal to the range of the second loop and can be much smaller (in the above calculations it is only 150 MHz), due to which the noise of the VCO-1 can be significantly reduced. These two factors contribute to improved spectral purity and agility of the synthesizer.

However, in order to obtain a smaller frequency step size, it is required to increase the division factors of all three DIVRs and to decrease the value of the reference frequency F_r, which, naturally, leads to a deterioration in the spectral purity of the signal and to a decrease in agility. The problem can be solved by increasing the cascades in this structure, that is adding one or more PLLs, but this is not always acceptable due to the increasing complexity.

Another possible way is to use fractional frequency dividers including the fractional-N PLL option. An example of such a solution is the cascade connection of two HMC830 microchips from Hittite (now part of Analog Devices). This microchip is the fractional-N PLL synthesizer with an integrated VCO.

An important feature of the structure under consideration is the following. With a sufficiently large set of division ratios, including their fractional values, practically the same output frequency can be obtained with different combinations of these ratios. This makes it possible to use the most successful combinations of them to get rid of spurious spectral components, for example, such as integer boundary spurs (IBS). This is when the VCO frequency is closest to one of the harmonics of the reference frequency.

1.6 THREE-LOOP SCHEME

Among the more complex structures of frequency synthesizers, the most widespread, perhaps, is the three-loop scheme shown in Figure 1.6. In it, two loops can be functionally distinguished, with fine and coarse frequency resolution, and the third one distinguished with a summing loop. In Figure 1.6, the names of the blocks of the fine and coarse loops are marked with the index 1 and the index 2, respectively.

The reference frequencies F_{01} and F_{02} (the comparison frequencies in the relative PDs) are obtained from a common reference frequency F_r using a frequency divider F_D with a fixed division factor.

The output of the fine loop forming the signal F_{SS} with a small step size is completed with a frequency divider with a fixed coefficient M due to which the noise of the F_{SS} signal at the input of the subsequent PD of the summing PLL is reduced by $20 \lg M$. The parameters of the fine loop are chosen in such a way as to obtain the smallest possible step size $dF_{SS} = F_{01}/M$ at the highest possible comparison

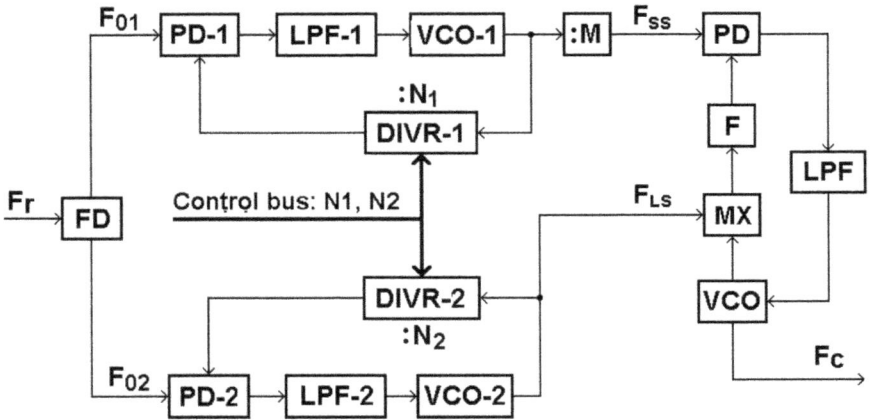

Figure 1.6   Three-loop scheme.

frequency F_{01}. The F_{SS} frequency range obtained in this case can be small, much less than the required F_c frequency range at the synthesizer output.

In the second loop, a large step size dF_{LS} is formed, equal to this limited range of F_{SS}, that is, $dF_{LS} = F_{SS}$. The summation of the step sizes, large and small, occurs in a summing loop. To do this, a frequency mixer MX is included in it, in which the F_{LS} frequency is subtracted from the F_c frequency or vice versa. The result of the subtraction is filtered by the filter F, which can be either a bandpass or a lowpass filter. The signal of the difference frequency from the filter output is fed to the second input of the PD for phase comparison with the F_{SS} signal. Thus, the coarse step size is filled with a fine one, and, as a result, the expression for the F_c frequency at the synthesizer output can be written as

$$F_c = F_{LS} \pm F_{ss} = N_2 F_{02} \pm N_1 F_{01}/M$$

Oscillators VCO and VCO-2 have almost the same frequency tuning range since $F_c \gg F_{SS}$, that is, the ranges differ only by a small amount of F_{SS}. Therefore, it is necessary to carefully match them in terms of control voltages in order to avoid "mirror" tuning of the output generator VCO.

In synthesizers, according to this structure, the output noise is determined in the band of the summing loop by the VCO-2 and behind the band by the VCO.

As an example, let's assume that it is required to obtain a frequency range from 700 MHz to 1 GHz with a step $dF = 10$ kHz. Then the following parameters of the PLLs can be selected: $F_{01} = F_{02} = 1$ MHz; change in the coefficient N_1 from 900 to 1,000 units; the tuning range of the VCO-1 is from 900 MHz to 1 GHz and $M = 100$, that is, the F_{SS} frequency range at the input of the summing loop is $F_{SS} = 9 \div 10$ MHz with a step of 10 kHz; change in the coefficient N_2 of the second loop from 691 to 990 MHz through one unit; and the tuning range of the VCO-2 is from 691 to 990 MHz with a step of 1 MHz.

The choice of such a rather high comparison frequency in PDs as 1 MHz makes it possible to provide a frequency switching time of the order of fractions of a millisecond with a high spectral purity of the output signal.

The disadvantage is the obvious complexity of the system, the need for a careful design with shielding of blocks in order to avoid combinational interference in the signal spectrum.

1.7 THROWER'S SCHEME

A variation of the three-loop synthesizer is the one proposed by Thrower [6]. It is shown in Figure 1.7. It uses two synchronously controlled dividers DIVR-2 and DIVR-3 with the same division ratios N_2 in two PLLs connected in series. In some periodicals, this scheme was called "twins."

In the first loop of the presented scheme, there is formed initial frequency step sized $F = F_{01}/M$. In the third one, a coarse step is F_r/M, and the second loop performs the summation function, but, unlike the summing loop of the previous scheme, the loop requirements are much weaker here since it performs practically on the same frequency F_r. The presence of a divider with a division factor M and a small tuning range of the VCO-2 reduce the noise requirements for the generators of the first and second loops. The output noise level in this scheme is determined primarily by the VCO-3.

According to the structure of the scheme, the output frequency F_c of the synthesizer is determined by the expression

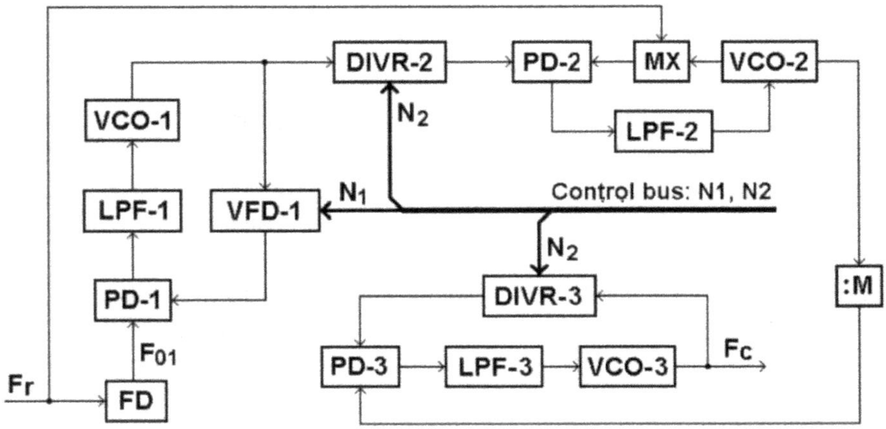

Figure 1.7 Thrower's scheme.

$$F_c = \left(N_1 F_{01} + N_2 F_r \right) / M$$

where F_{01} is a frequency after divider F_D.

Similarly to the previous scheme, this structure allows use of higher comparison frequencies in the loops, which makes it possible to implement a synthesizer with improved spectral purity and agility.

The disadvantages of three-loop synthesizers are quite obvious:

- A significant increase in size, power consumption, and cost in comparison with a single-loop synthesizer;
- The presence of three high-frequency VCOs is, as a rule, the cause of the formation of interference components in the spectrum of the synthesizer for elimination of which the additional efforts of the circuit-constructive plan are required.

References

[1] Young, C. J., "Stabilized Oscillator Generator," US Patent 2,490,500, 06.12.1949, filed December 29, 1946.
[2] Woodward, J. D., "Variable Frequency Oscillation Generator," US Patent 2,490,499, 06.12.1949, filed April 23, 1947.
[3] Tollefson, R. D., "Frequency Synthesizer," US Patent 3,435,367, 28.06.1971, filed January 16, 1969.

[4]   Denisov, G. V., "Method for Obtaining a Discrete Frequences," Patent of Russia No. 171025, priority 03.02.1958, in Russian.

[5]   Martin, D. J., "Frequency Synthesizers," US Patent 3,600,683, 17.08.1971, filed June 20, 1969, Priority GB 30796/68.

[6]   Thrower, K. R., GB Patent 1,303,631, 17.01.1973, filed March 1, 1969.

2

OTHER SCHEMES

2.1 CHENAKIN'S SCHEME

Frequency synthesis idea proposed by Alexander Chenakin was disclosed in his American patent [1] and described in a number of publications, for example, [2–4]. Here we will restrict ourselves to a description of a simplified scheme in order only to clarify the idea itself without going into too much detail.

The scheme is shown in Figure 2.1. It is a PLL system, which, as usual, contains a PD, a lowpass filter, and a VCO. The peculiarity of the synthesizer is building a feedback of the system.

This circuit contains two parallel paths operating in series in time and forming two PLLs. When a new value of the synthesized frequency is set, the key switches to the appropriate position so that the first loop (the one in which the frequency divider with a variable division factor N is turned on) will work and bring the signal frequency to the specified value with an accuracy sufficient to lock the frequency in the second loop. The output of the last stage of the reference shaper also serves as a reference one for both PLLs.

After that, the second loop is turned on. It operates via a multistage frequency converter in a feedback loop. The frequency converter contains series-connected frequency mixers, in which, as a result of interaction with the reference frequencies, the signal frequency is shifted down to the comparison frequency in the PD. The reference

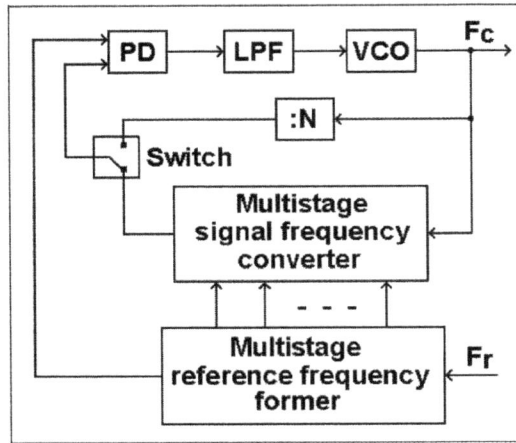

Figure 2.1 Chenakin's scheme.

frequency for each subsequent stage is lowered using a multistage reference frequency converter, consisting of a set of dividers, multipliers, and frequency mixers. This unit operates from a comparatively high-frequency reference source of frequency F_r. There, it can be said, is an analogy with a conventional analog synthesizer with the only difference that the frequency conversion is downward, not upward. The output of the last stage of the former reference also serves as a reference for both PLLs.

The task for the frequency converter is to bring the spectrum of the controlled generator to the PD without dividing the frequency. Due to this, the gain in this loop is not reduced, as it is in the first loop, and this achieves a high efficiency of suppression of the noise of the controlled oscillator.

It is important to note that the intermodulation products of such a multistage frequency converter are harmonics of the PD reference frequency and can be easily suppressed by the PLL filter. Thus, the proposed architectural solution makes it possible to obtain rather low side components (spurs) in comparison with classical offset schemes.

However, the noted advantage of the scheme is realized only with a sufficiently large frequency step that is equal to the comparison frequency in the PD. With a decrease in the step, the comparison frequency also decreases, the loop bandwidth narrows, and, accordingly, the quality of the signal spectrum and the speed of the synthesizer decrease.

The described scheme can be used as part of a more complex structure additionally including means for filling coarse steps with finer ones. For example, another PLL and DDS can be added for such a summing. In this case, there is a need to take measures to ensure a sufficiently low level of noise and spurs from DDS.

As a result, the overall structure of the synthesizer becomes much more complicated. It has been implemented in practice in the Phase Matrix (now part of National Instruments). It is a line of synthesizers such as QuickSyn (see the same sources cited above) where it showed very high spectral purity characteristics. In the QuickSyn Lite FSW-0010 model, which provides the range from 0.5 to 10 GHz, the noise floor at a signal frequency of about 10 GHz in the PLL bandwidth is about −120 dBc/Hz, the level of spurs is less than −70 dBc, and the switching time is of the order of tens of microseconds.

In the next model, FSW-0020, the frequency range is expanded to 20 GHz, and a number of millimeter-wave synthesizers have been created on its basis, the characteristics of which are shown in Table 2.1. In this case, in all the above modifications, the frequency step size is equal to 1 Hz, the spurs do not exceed −60 dBc, and the frequency switching time is not more than 1 ms.

The disadvantage is the high enough complexity and, accordingly, the cost. At the same time, everything is relative. If to take into account the rather high ratio of quality to complexity and cost then this can hardly be attributed to disadvantages. Moreover, the obtained characteristics of spectral purity and speed are not architectural limitations in principle, but a specific and very simplified implementation. These characteristics can be significantly improved by more fully exploiting the potential of this synthesis method.

Table 2.1

Model	Band	Phase Noise
FLS-2740	27–40 GHz	−103 dBc/Hz at 40 GHz
FLS-5067	50–67 GHz	−100 dBc/Hz at 67 GHz
FLS-7682	76–82 GHz	−98 dBc/Hz at 82 GHz

2.2 GOREVOY'S SYNTHESIZER

Not all customers need ultralow noise and spurs, as well as extra high agility that in the sum is reached, as a rule, at the expense of increase in power consumption, dimensions, and cost. Therefore, there is a problem of development of such a device in which characteristics in the first group of requirements can be a little weakened, but in the second group they are considerably toughened. One of the versions of the solution of this task is presented in [5–7].

The proposed scheme of a synthesizer is explained with Figure 2.2, which illustrates how to get from the reference frequency F_r = 100 MHz multiplied by 20 times a signal F_c in the range from 25 to 6,000 MHz.

The synthesizer consists of two blocks connected in series: the reference, and the main synthesizers. The first of them provides frequency with a small step size in rather small range, and the second one uses this frequency as the reference one expanding the range up to several octaves. Frequency transformations, as well as designations of elements in the scheme, do not demand explanations.

Actually, it is the known principle of creation of a synthesizer, which, in particular, is used in QuickSyn (see the previous section).

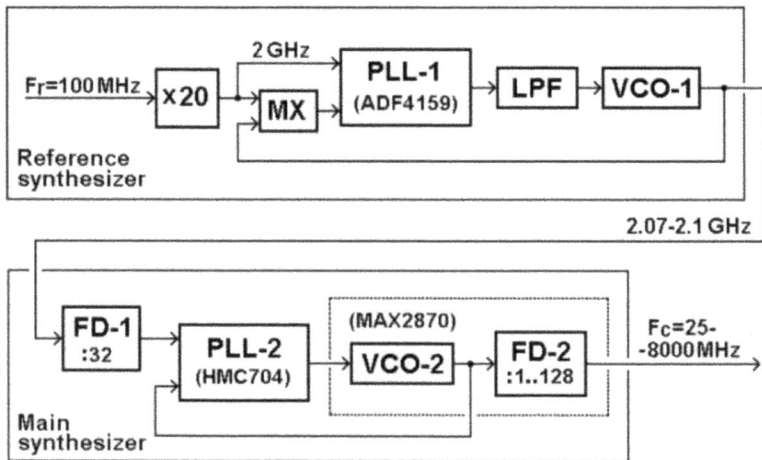

Figure 2.2 Gorevoy's synthesizer.

Novelty consists in the successful selection of chips for the called blocks.

For example, instead of DDS for getting a small step size, as it takes place in the QuickSyn, in a reference synthesizer, the fractional frequency divider with a delta sigma modulator is used. For this purpose, the ADF4159 chip from Analog Devices approaches, and this significantly saves power consumption and dimensions because the divider is in the structure of a chip (PLL-1), where other necessary circuits are also placed (e.g., the phase-frequency detector and current switch). Actually, the system consumes so much, how many would it consume without DDS (economy about 0.5 to 1W)? The other characteristics do not strongly concede to the variant with DDS. Also, in the block of the main synthesizer, inexpensive serial chips can be used: HMC704 from Hittite (PLL-2) and MAX2870 from Maxim Integrated (VCO-2 and FD-2).

According to the described structure, a company "Mikran" (Tomsk, Russia) has developed a portable USB synthesizer PLG06 [8] with the following main characteristics: output frequency range of 25 to 6,000 MHz with a step size of 1 Hz; phase noise of −122 dBc/Hz at offset 10 kHz from a carrier 1 GHz; spurs of −70 dBc; harmonics of −30 dBc; and frequency switching time of 100 μs. There are analog modulation modes: AM, PM, FM, IM (external/internal source), and scan mode. The device, having the functionality of classic laboratory generators, consumes only 2.5W and is powered and controlled via a single USB 2.0. The dimensions of the device are only 125 × 65 × 25 mm.

2.3 SCHEME WITH DDS IN PLL

Let's consider the idea of DDS in a PLL with the example of an SG8-HP01M generator from a company Advantex, Moscow [9]. The generator scheme is shown in Figure 2.3.

The DDS in this scheme acts as a variable division frequency divider. It is clocked by 8 times the divided frequency of an octave range of 4 to 8 GHz from the VCO. The division factor of the DDS, as a frequency divider, is also tuned in the octave range in accordance with the octave frequency range of 0.5 to 1 GHz at its input. Thus, the frequency at the DDS output remains constant and equal to the

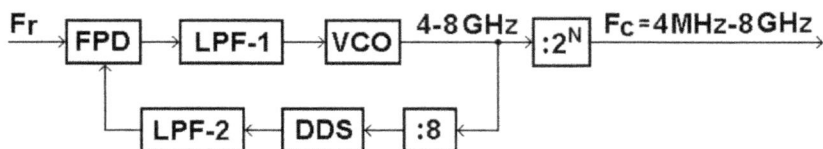

Figure 2.3 An example of a synthesizer with DDS in PLL.

reference frequency F_r (in the presence of synchronism in the loop). The largest octave range is obtained directly at the output of the VCO, and the other, smaller ones, after dividing the upper range with binary frequency dividers, will be discussed in Section 2.5.

The expression for the frequency F_c at the synthesizer output can be written as:

$$F_c = 2^{21} F_r / (KM)$$

where the coefficient $K = 2^N$ of the output frequency divider takes values from 1 to 2^{10}, M is a specified integer in the range of values from M_{min} to M_{max}, and

$$M_{min} = 2^{18} F_r / 10^9 \text{ and } M_{max} = 2M_{min}$$

The obvious advantage of the scheme, like any other one-loop structure, is the simplicity of its implementation. The disadvantage is a rather high level of nonharmonic components in the signal spectrum originating from the DDS. With a constant value of the reference frequency F_r, it reaches −50 dBc in the 2-MHz offset band. This level can be reduced by 10 dB using two reference frequencies that are automatically switched by the built-in program. The phase noise is −120 dBc at a signal frequency of 1 GHz with an offset of 10 kHz, and the frequency switching time is about 4 ms.

2.4 SADOWSKI'S SCHEME

The idea proposed by the author is distinguished by a special way of constructing a fractional divider included in a PLL [10, 11]. A frequency synthesizer scheme that uses this idea is shown in Figure 2.4. A fractional frequency divider is represented in it by two dividers with integer coefficients K and L and a frequency mixer MX with a filter F at its output. The resulting division factor for such a structure is

$$N = KL/(K \pm L)$$

where $K > L$.

The advantage of the idea is that such a frequency divider, having the properties of fractional division, does not have fractional components at its output. This can be shown with an example.

Suppose that with the reference frequency $F_r = 10$ MHz, it is required to obtain the frequency $F_c = 119$ MHz at the output of the VCO included into the PLL. In this case, it is necessary to have a division ratio in the loop equal to $N = 11.9$. It can be ensured by setting the following values for the division coefficients: $K = 17$ and $L = 7$. Then the frequencies at the inputs of the mixer MX will be equal, respectively, $F_K = 119/17 = 7$ MHz and $F_L = 119/7 = 17$ MHz, and their difference at the mixer output will be 10 MHz, which is used for comparison with the 10-MHz reference frequency in the PD.

If the summation of the frequencies F_K and F_L are used with the same division factors 17 and 7, then the resulting division factor will be obtained

$$N = (17 \times 7)/(17 + 7) = 4.9583(3)$$

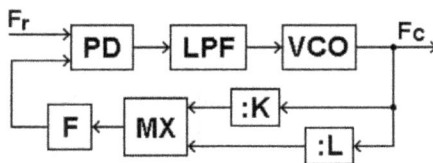

Figure 2.4 Sadowski's scheme.

and the corresponding frequency is F_c = 49.583 (3) MHz. In this case, the frequencies at the outputs of the corresponding dividers are F_K = 2.916(6) MHz and F_L = 7.083 (3) MHz, and their sum is 10 MHz, which, as in the previous case, is used for comparison in the phase detector.

The disadvantage of this structure is the need to use the filter F in order to get rid of combinations like $+/-nF_K+/-mF_L$. This significantly limits the possibilities of a wide choice of coefficients K and L. In addition, the desire to provide high-frequency resolution leads to the need for a corresponding increase in these coefficients and a narrowing of the filter bandwidth, which accordingly reduces the synthesizer speed.

The disadvantages also include a relatively complex algorithm for selecting the required signal frequency. For each specific requirement of the synthesizer frequency range, frequency step size, and speed, a table with precalculated values of the K and L coefficients is required.

However, due to the indisputable advantages of the scheme in comparison with others using both integer and fractional division factors, this scheme could find practical application, albeit limited by the noted disadvantages. The scheme is analyzed in more detail in [12].

2.5   EXTENDING THE FREQUENCY RANGE

It is clear that the limits of VCO tuning are limited, primarily due to the need to ensure an acceptable noise level. Oscillators with overlapping frequencies of more than an octave are practically not used. You can expand the frequency range of the synthesizer using a set of several switched oscillators, but it is quite difficult and expensive. However, if the octave range has already been obtained, then you can further expand the frequency range of the synthesizer downward in frequency in a fairly simple way, as shown in Figure 2.5.

Figure 2.5 shows an example of obtaining the frequency range $F_c = F_{c0}/2$, where F_{c0} is the original octave range. To avoid falling into the spectrum of the signal F_c of its subharmonics, the subsequent frequency dividers, after the used ones, are turned off. As frequency dividers, it is advisable to use triggers with an output voltage in the

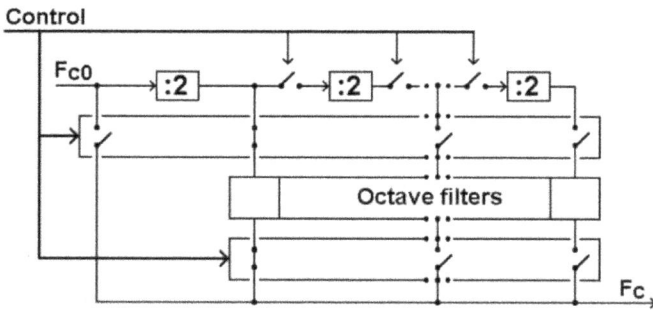

Figure 2.5 Scheme for extending the frequency range.

form of a square wave that does not contain the second harmonic. Therefore, a sinusoidal signal at the output of F_c can be obtained using fairly simple lowpass filters of the octave range.

This method of extending the frequency range is widely used in practice [13], in particular, in the developments of Phase Matrix/NI, USA (FSW-0010), Stanford Research Systems, USA (7SG392, 7SG394, 7SG396), AnaPicoInc, Switzerland (APSIN6010), Advantex, Moscow (SG8), Mikran, Tomsk (PLG06), and some other companies.

References

[1] Chenakin, A., "Low Phase Noise PLL Synthesizer," US Patent 2009/0309665 A1, December 17, 2009, filed September 5, 2008.

[2] Chenakin, A., "Novel Approach Yields Fast, Clean Synthesizers," *Microwaves & RF*, October 2008.

[3] "An Interview with Alexander Chenakin," *Microwaves & RF*, March 2009.

[4] Chenakin, A., *Frequency Synthesizers: From Concept to Product*, Norwood, MA: Artech House, 2010.

[5] Gorevoy A. V., "Frequency Synthesizer," Patent of Russia No. 2523188, priority 09.04.2013, in Russian.

[6] Gorevoy, A. V., "Obtaining Subhertz Resolution in Frequency Synthesizers with a High Degree of Spectral Purity and Low Power Consumption," *2014 24th Intl. Crimean Conference Microwave & Telecommunication Technology (CriMiCo'2014)*, September 7–13, 2014, Sevastopol, Crimea, Russia, in Russian.

[7] Portable USB synthesizer, http://www.micran.ru/sites/micran_ru/tmpl/micran_ru/inc/pdf/PLG06.pdf.

[8] Signal generator SG8-HP01M, SG8-HPSS01M, Specifications, http://ad5antex-rf.com/Downloads/SG8_Manual_en.pdf, in Russian.

[9]   Sadowski, B., "A Self-Offset Phase-Locked Loop," *Microwave Journal*, April 2008.

[10] Sadowski, B., "A Fractional Radio Frequency Multiplier," *MIKON 2008 – 17th International Conference*, 2008.

[11] Nikitin, Y. A., "Building a Path for Bringing an Active Frequency Synthesizer," *Izvestiyavuzov, Instrument Making*, Vol. 55, No. 3, 2012, pp. 19–26, in Russian.

[12] Kuzmenkov, A. S., A. E. Polyakov, and L. V. Strygin, "Overview Analysis of Modern Architectures of Frequency Synthesizers with PLL," *MPTI PROCEDURES - Radio Engineering and Telecommunications*, Vol. 5, No. 3, 2013, pp. 121–133, in Russian.

[13] Braymer, N. B., "Frequency Synthesizer," US Patent # 3,555,446, 12.01.1971, filed January 17, 1969.

3

SCHEMES WITH THE FRACTIONAL DIVIDER AND WITH THE SUPPRESSION OF FRACTIONAL NOISE AND OTHER SCHEMES

Let's move on to considering single-loop structures with a fractional frequency divider and various circuit options for the suppression of fractional noise.

3.1 BRAYMER-GILLETTE SCHEME

One of these options is shown in Figure 3.1. In general terms, it was almost simultaneously patented by Braymer [1] and Gillette [2]. True, in the descriptions of their patents, much attention was paid to the construction of the original schemes of fractional dividers, although this does not apply to the principle of suppression of fractional noise and therefore is not reflected in the figure.

The fractional divider is presented in the form of an integer part with a division factor N_0 and a fractional part performed on an accumulator. The overflow pulse of the latter is transmitted to the integer part, and the overall division factor increases by one, which is why the fractional interference occurs.

To delete it, a digital-to-analog converter (DAC) is used, with the help of which a signal is formed, a copy of the interference detected in the PD. In the summer, the voltages from the DAC and PD outputs

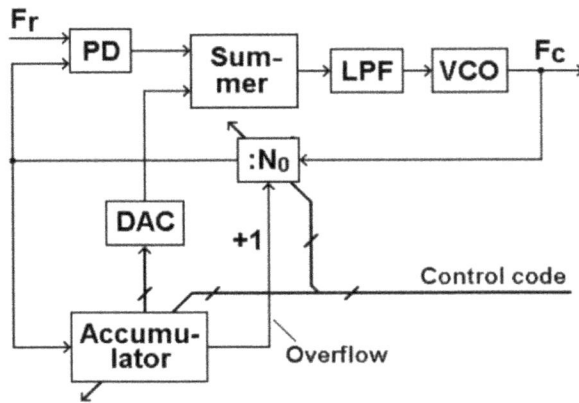

Figure 3.1 Braymer-Gillette's scheme.

are added in antiphase, which causes the fractional interference to be suppressed.

It is clear that the degree of interference suppression depends on the accuracy of the DAC and on the accuracy of the summer, and these accuracies are naturally limited. Therefore, if the capacity of the accumulator is quite large in order to get a sufficiently small step size of the frequency, it makes no sense to take a DAC of the same capacity, it is limited to 12 to 14 digits connected to the corresponding most significant bits (MSBs) of the accumulator.

3.2 INTEGRATOR OPTION

Another version [3] of the fractional noise suppression scheme is shown in Figure 3.2. It uses basically the same blocks as in the previous scheme, only an integrator has been added and the DAC is for a different purpose.

Each accumulator overflow causes a frequency jump at the output of the frequency divider. To suppress it, a signal corresponding to the resulting phase deviation needs to be created and added in antiphase with the voltage at the output of the PD. This is what the integrator is for. It can be based on an operational amplifier.

The required level of the suppression signal is inversely proportional to the division ratio. Therefore, with a sufficiently large frequency overlap of the synthesizer and a wide range of changes in the

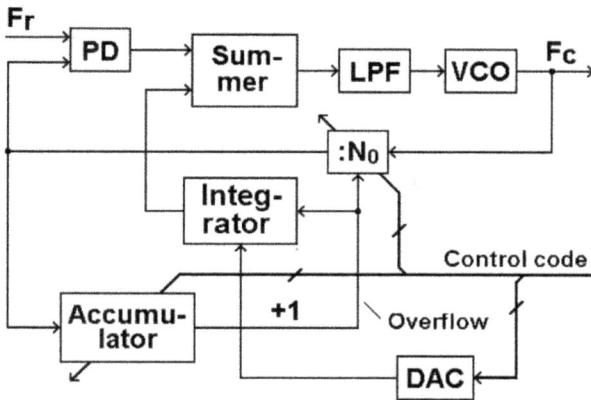

Figure 3.2 Option of a scheme with an integrator.

division ratio, it is necessary to control the integrator gain for which the DAC is used. It can be used as a source to power the integrator so that the voltage at its output is controlled by a control code. In the previous scheme, this possibility was absent. However, even there, an additional DAC can be entered that powers the existing one (which must be of the multiplying type) and is connected to the control bus.

3.3 COX'S SCHEME

By its structure, Cox's scheme [4] can be attributed to direct digital synthesizers with some peculiarities. The signal in it is obtained by dividing the reference frequency by a variable fractional ratio followed by the suppression of fractional interference by means of a programmable time shift of the signal at the output of the scheme.

The scheme is shown in Figure 3.3. It contains a programming device (Programmer) for setting the integer N_0 and fractional n parts of the N coefficient, an absorbing counter (indicated in the scheme as N), an accumulator for forming the fractional part of the division coefficient, and a delay generator controlled by a DAC. The absorbing counter is clocked by the pulses of the reference frequency F_r, and the accumulator is clocked by the signal pulses of the frequency F_c. The absorbing counter, together with the accumulator, forms, as a whole, the controlled fractional divider.

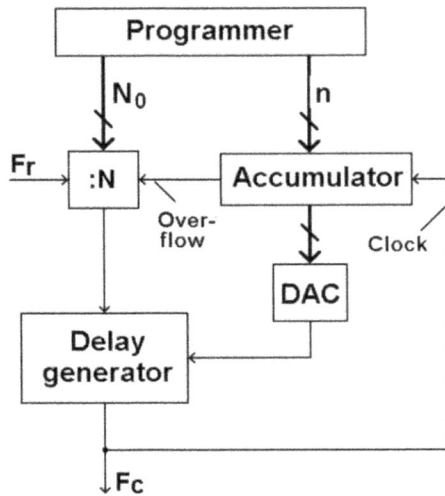

Figure 3.3 Cox's scheme.

The operation of the scheme can be seen on a specific example. Let's assume that, with the reference frequency equal to $F_r = 100$ MHz, it is required to obtain the signal frequency $F_c = 30$ MHz. This means that at a certain frequency resolution, a pulse at the synthesizer output should appear every 3.3333 reference pulse periods. For this, the integer part of the division factor N is set equal to $N_0 = 3$, and, to ensure the fractional part of this factor, the number at the input of the accumulator is 3,333, with its capacity equal to 10,000. It is clear that, in this case, the sought frequency of 30 MHz will be obtained with an error of 300 Hz. Let's also assume that the absorbing counter and the accumulator are both included in operation at zero initial conditions.

The absorbing counter is designed in such a way that one reference pulse is deleted from its input by the accumulator overflow pulse.

The first three reference pulses pass freely through the absorbing counter, creating the first signal pulse, which then passes unhindered through the delay generator to the output of the circuit. This is because, first, there is no accumulator overflow pulse, and second, the accumulator content, like the DAC, is zero and, therefore, the delay generator does not create a time shift for this pulse. This impulse, acting on the accumulator, changes its content from 0 to 3,333.

The next 3 reference pulses also pass unhindered through the absorbing counter, creating a second signal pulse at its output. However, then this pulse passes to the output of the circuit with a delay of 0.3333 of the period T_r of the reference pulses, which is created by the delay generator under the influence of the signal from the DAC output. This signal pulse increases the accumulator content to a value of 6,666.

After three subsequent pulses F_r, a third signal pulse is obtained at the output of the absorbing counter, which passes to the output of the scheme with a delay of 0.6666 T_r in accordance with the new value of the accumulator content. The fourth signal pulse with a delay of 0.9999 T_r is generated in the same way.

At the fifth signal pulse, the accumulator overflows, its content is reset to 3,332, and its overflow pulse deletes one F_r pulse at the input of the absorbing counter. Then the scheme operates according to the described algorithm, aligning, with the help of a delay generator, the arrangement of signal pulses in time to make the process periodic, that is, to eliminate the noise of fractionality.

One of the possible options for the delay generator circuit is shown in Figure 3.4. In the absence of a pulse from the divider output, the switch is closed, which prevents the capacitor C from being charged from the current source.

At the same time, the trigger is in the 0 state. With the appearance of the said pulse, the switch opens, and the current source charges the capacitor C according to a linear law. The voltage from the capacitor

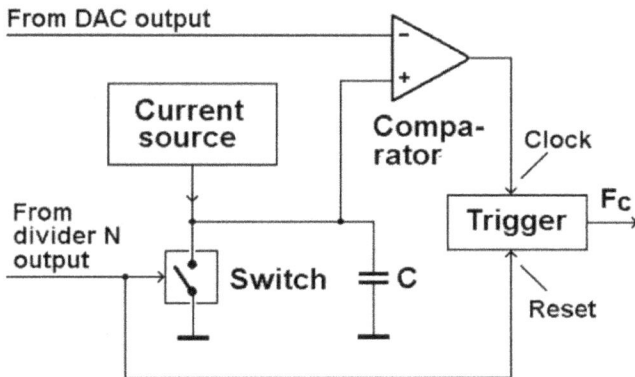

Figure 3.4 Delay generator circuit.

is compared in the comparator with the voltage at the output of the DAC, and, when they are equal, a pulse appears at the output of the comparator, which transfers the trigger to the state 1. The time interval between pulses from the divider and trigger outputs is a linear function of the voltage from the DAC output. The circuit parameters are calculated in such a way that the maximum voltage from the DAC output corresponds to a delay equal to one pulse period of the reference frequency F_r.

In terms of the effectiveness of the action, the schemes considered above are approximately equivalent. Due to the relatively low accuracy of DAC, summation, and analog integration, it is not possible to achieve high spectral purity of the signal in them, which limits the area of their use.

3.4 UNDERWOOD'S SCHEME

The scheme is shown in Figure 3.5 [5]. It uses an accumulator to divide the reference frequency F_r with the fractional coefficient $N = Q/A$, where Q is the capacity of the accumulator, and A is the accumulated number contained in the control code N. Thus, an output of the accumulator is a reference for the PLL, which forms the signal frequency F_c. The other input of the PD is connected to the output of the VCO.

As a result, the frequency F_c is equal to the average frequency of pulses at the output of the accumulator, that is,

$$F_c = F_r/N = AF_r/Q$$

A DAC is used here to suppress the fractional noise. In it, the remnants from the overflow of the accumulator are converted to an

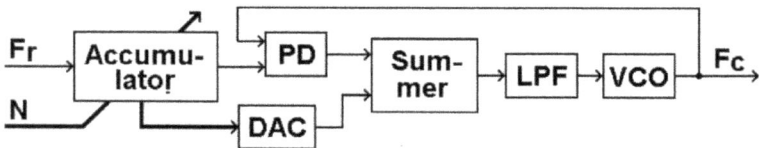

Figure 3.5 Underwood's scheme.

analog form, and this process is added in the summer in antiphase and with a corresponding weight with the output signal of the PD, which eliminated the fractional interference. The lowpass filter serves to suppress the components with the frequency F_c and their harmonics, as well as the residuals of the high-frequency components of the fractional interference.

The level of fractional noise at the synthesizer output depends both on the accuracy of the DAC and on the accuracy of the summation of signals in the summer. True, the weight ratios of the added signals are constant, and this makes it easier to achieve a higher suppression than in the Breimer-Gillette scheme and in the scheme with an integrator discussed above.

It is also important to note that the signal frequency F_c is lower than the reference frequency F_r. To raise the frequency range of the signal up, there can be additionally included a frequency divider with ratio M in the signal path. However, at the same time, naturally, the fractional interference in the signal spectrum will be increased by 20lgM dB.

Such an option is also possible, when the accumulator, as a divider, is included in the signal path. Then means will be required to control the weight ratios in the summer, as was in the case with the integrator and in the Braimer-Gillette scheme. This will inevitably lead to a decrease in the total suppression accuracy, that is, to a deterioration in the spectral purity of the signal.

3.5 REVISION OF THE UNDERWOOD SCHEME

The accuracy of summing signals in the adder can be significantly increased, and therefore the spectrum can be improved, by replacing the analog summation with a digital analog, as shown in Figure 3.6 [6].

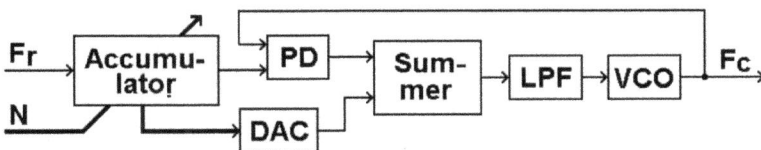

Figure 3.6 Revision of the Underwood scheme.

The operation of the circuit is explained in Figure 3.7, where an RS trigger is used as the PD. For example, $Q = 8$ and $R = 3$ are selected. As can be seen from Figure 3.7, the resulting process in the DAC comprises a constant E_c component used to control the VCO frequency and two high-frequency sawtooth components H and G with respective frequencies F_r and F_c eliminated by the lowpass filter. There is no low-frequency interference, although it is contained in both the highest bit (pulse width modulation) and the lowest bits (process irregularity). This is due to the fact that the interference in the highest bit is compensated just by the corresponding irregularity of the process in the lowest bits. Interference of this kind can be called fractional interference, since it occurs due to the mutual noncrateness of the numbers Q and R. Multiplication in the PLL is absent.

The operation of the circuit is illustrated using Figure 3.7, where an RS flip-flop is used as a PD. For example, the values $Q = 8$ and

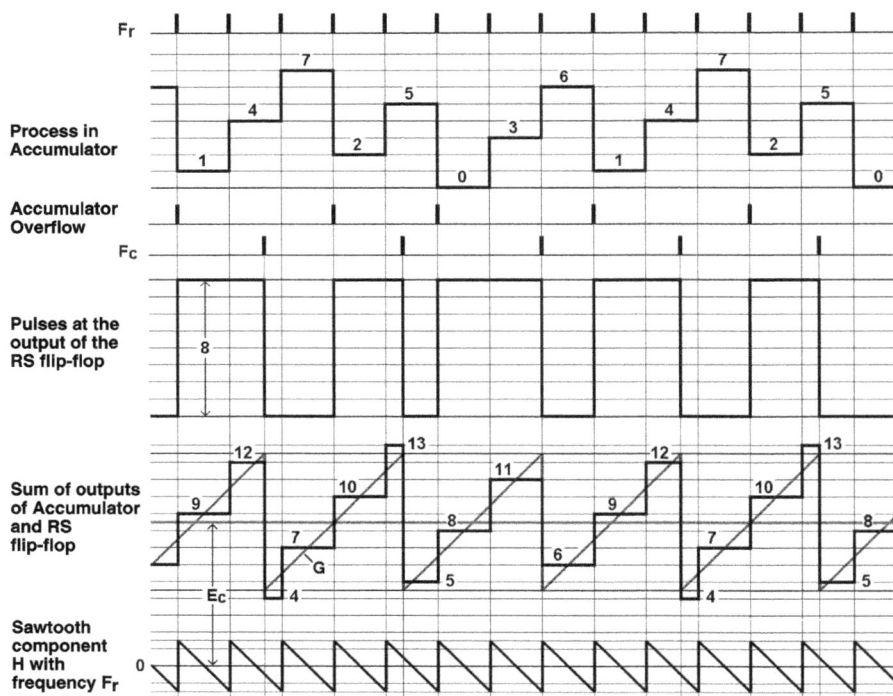

Figure 3.7 Diagrams explaining the operation of the diagram in Figure 3.6.

$R = 3$ are chosen. As can be seen from Figure 3.7, the resulting DAC process contains a direct current (DC) component, E_c, used to control the VCO frequency, and two high-frequency sawtooth components, H and G, with corresponding frequencies, F_r and F_c, removed by a low-pass filter. There are no low-frequency interferences, although they are contained both in the most significant digit (pulse-width modulation) and in the lower digits (irregularity of the process). This is explained by the fact that interference in the most significant bit is compensated by the corresponding irregularity of the process in the least significant bits (LSBs). Noise of this kind can be called fractional noise, since they arise due to the mutual nonmultiplicity of the numbers Q and R. There is no multiplication in the PLL.

Naturally, the quality of the signal spectrum is most affected by the MSB inaccuracy. It is known from Analog Devices that the inaccuracy of one segment of a DAC can be reduced to a value not exceeding 0.1% of its weight. This fact can also be attributed to the MSB category of the scheme in question.

Using a fast Fourier transform (FFT) apparatus, the signal spectrum at the output of the MSB discharge (RS flip-flop) can be calculated. As an example, this is done and shown in Figure 3.8 for the case of $Q = 64$, $R = 5$. There is a diagram of the process, its spectrum normalized by the level $Q = 64$, as well as a graphical interface of the program for reading the entered parameters. In this case, it should

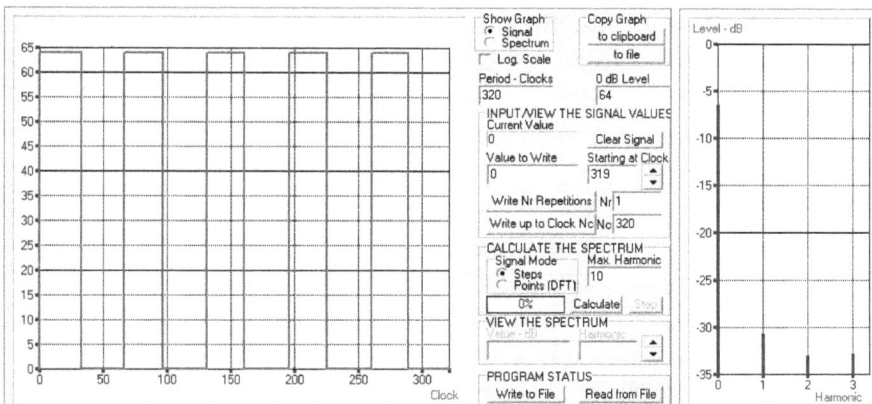

Figure 3.8 Calculating the spectrum at inaccurate MSB.

be borne in mind that the weight of the most significant discharge is one-half of the full voltage scale at the output of the PD, so that the spectrum result should be reduced by 6 dB.

It is understood that with 0.1% inaccuracy of the MSB discharge, interference occurs in the spectrum in the form of an exact copy of the calculated spectrum reduced by 60 dB. Further, it should be noted that, at the output of the PLL, it will split into two sidebands, reduced by 6 dB and increased by 16 dB in accordance with the slope of the PD characteristic equal to $E_{max}/(2\pi)$, where E_{max} is the full control voltage scale for the VCO. As a result, when converting the interference (on the graph, it is the first harmonic with a level of −31 dB) to the output of the PLL, we get: − 31 − 6 (normalization) − 6 (two sidebands) + 16 (slope PD) − 60 (percent inaccuracy) = −87 dB.

Similarly, interference due to the total inaccuracy of the LSBs can be calculated. The corresponding calculations are shown in Figure 3.4.

Here, the interference level is less, but insignificant, by only 1 dB, which can be neglected. Therefore, the results obtained in both cases can be considered a consequence of a mismatch in the conjugation of the senior category with the younger ones.

Table 3.1 shows the results of calculations at different ratios of reference frequency F_r to frequency of signal F_c.

A simple pattern is seen from Table 3.1: The maximum interference level decreases in proportion to the F_c/F_r ratio. So it is possible to extrapolate and interpolate with a variation of the F_c/F_r ratio. There may be other factors affecting the spectral purity of the signal, but the results obtained are high enough that these unknown factors do not cause them to decline to a level unsuitable for practice.

Table 3.1
Interference Level Versus the F_r/F_c Ratio

Ratio F_r/F_c	64/21	64/11	64/5	64/3
Maximum Interference Level, dBn	−73	−81	−87	−91

It is also clear that the inputs for the signals F_r and F_c in Figure 3.1 can be swapped, and then, in the PLL system, there will be a multiplication of the reference frequency in F_c/F_r times with the corresponding increase in the interference level by $20\lg (F_c/F_r)$. This option may be quite acceptable in practice, since it does not require an increase in the reference frequency to raise the signal frequency.

3.5.1 Comparison with Other Synthesis Methods

As follows from the above analysis, the frequency synthesis method is based on two technologies: direct digital synthesis (battery and DAC) and PLL (LPF and GUN). Therefore, for short, it can be called DS-PLL, where DS is direct or digital synthesis, and PLL is PLL. From the first technology, DS-PLL has a high spectral purity of the signal, and from the second, together with the first, it has a simplicity of implementation in the form of a single-mesh structure. In terms of speed, that is, the speed of switching to a new frequency, DS-PLL is inferior to DDS (due to PLL), but significantly exceeds it in terms of spectrum quality. For DDS, there is an insurmountable limit of -53 dBc, regardless of the F_r/F_c ratio and the type of synthesizer chip, as follows from the ADIsimDDS authorship program ADI.

When comparing with frequency division synthesizers in the PLL, it should be noted that in the single-loop version, due to the extremely low characteristics of spectral purity and speed, they are practically not used. They are used only with small division factors in complex, bulky, multichain structures with high power consumption and expensive to produce. True, there are inexpensive chips of synthesizers such as fractional-N PLL (with delta sigma modulation) on the market, but their characteristics are low and unacceptable for high-end equipment.

The modern technological base is quite ready for the production of DS-PLL synthesizers. Batteries and DACs are widely used in both low-cost, frequency-limited DDS and gigahertz-powered samples. Both have up to 16 bits in the DAC and up to 32 or more bits in the batteries.

In conclusion, the described frequency synthesis technique may be of interest to those skilled in the art for practical use.

3.5.2 Variant with Pulse PD of "Sampling-Storage" Type

The scheme of the variant is shown in Figure 3.6 [7]. It contains series connected accumulator, DAC, lowpass filter, and pulse detector of "sample-storage" type. Its work is explained with the help of Figure 3.7.

The accumulator is clocked by the reference frequency F_r. As an example, it contains 4 binary digits, and the code on its input is $R = 5$. The pulse, stepping process in the accumulator and, accordingly, the proportional voltage at the DAC output include two sawtooth components: nonperiodic G and periodic H with reference frequency F_r. The high-frequency component H is suppressed by a lowpass filter, and the low-frequency component G passes to the analog input of the sampling-storage type detector S-S.

At the pulse input of the detector, pulses of signal frequency F_c are fed sampling from the component G. This component has the property that under the condition of synchronism in the loop, the value of the samples, from pulse to pulse, remains constant, and, thus, an E_c voltage is generated to control the frequency of the VCO.

With its relative simplicity, the scheme has a significant drawback. To obtain the component G with the desired accuracy, it is possible only if the frequency ratio F_r/F_c is large enough to suppress the component H to the required degree, without introducing significant distortions into the component G. Otherwise, the level of fractional interference in the signal spectrum may be unacceptably high. Therefore, the scheme can be used in the range of rather low frequencies of the signal, when the sawtooth component G is not distorted by the lowpass filter in the upper part of its frequency range.

It is easy to see that this scheme can also be used as a direct-type synthesizer. If a threshold element is included after the filter, then, at its output, we get pulses of the synthesized frequency $F_c = RF_r/Q$ with the frequency step size equal to $dF = F_r/Q$, where Q is the accumulator capacity. At the same time, the disadvantage noted above remains in force.

3.6 NIKIFOROV'S SCHEME

The drawback of the previous version is eliminated in the scheme proposed by Nikiforov [8, 9] and shown in Figure 3.8. The diagrams explaining its operation are shown in Figure 3.9.

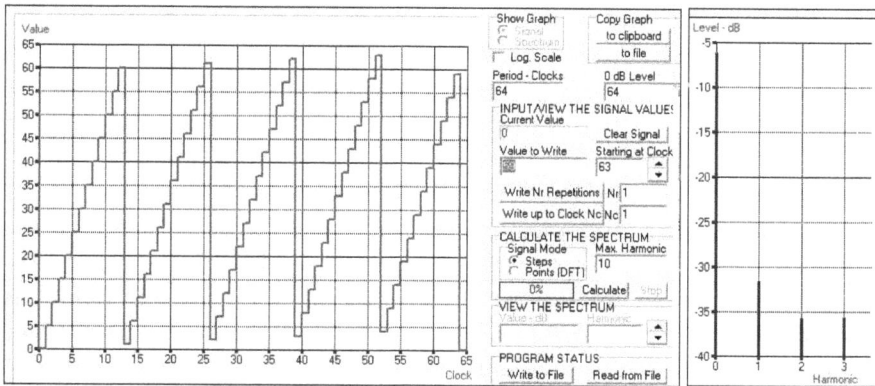

Figure 3.9 Calculating the spectrum at inaccurate LSB.

The accumulator is clocked by the reference frequency F_r. To avoid unnecessary complexity in the scheme description, we have chosen here small values of both its capacity $Q = 16$ and its accumulated number $R = 3$. The accumulator overflow pulse is fed to the pulse former synchronized with the frequency F_r. The digital sequence from the accumulator output is fed to one of the inputs of multiplexer, and the R code is fed to the other input of the latter.

The multiplexer is switched by a pulse from one of the outputs of the former pulse in such a way that the R code and the residue H in the accumulator as a result of its overflow alternate at the output of the multiplexer with each accumulator overflow. The duration of the remainder is doubled with respect to the period $T1 = 1/F_r$. Then the digital sequence from the output of the multiplexer is converted by the DAC into an analog equivalent and fed to the integrating element, which can be based on an operational amplifier. The discharge element serves to reset the charge in the integrating stage during an accumulator overflow. A control pulse from the other output of the pulse former is used for this purpose. The duration of this pulse is $T2 = 2/F_r$. During this time, the integrating circuit must be completely cleared of charge.

The diagrams in Figure 3.9 show: (a) the process in the accumulator, (b) a control pulse of the integrating circuit discharge, (c) a control pulse of the multiplexer, (d) the current code values at the multiplexer output and proportional to them analog values at the DAC output, and (e) the voltage at the integrating circuit output.

Attention should be paid to the characteristic features of the diagram E at the moments of time, marked by numbered points on the abscissa axis. At point 0, the integrating element is completely discharged. At point 1, the multiplexer has switched on the number $R = 3$ at the DAC input, and at the time interval up to point 2 the link is charged according to the linear law with the rate determined by the equivalent of this number at the DAC output. At point 2, the accumulator overflowed; the B pulse resets the charge of the integrating link; the C pulse turned on the residual from the accumulator overflow through the multiplexer. At point 3, the integrating circuit is charged at a rate proportional to the residual $H = 2$ converted into the DAC. At point 4, code $R = 3$ is enabled again, and the integrating circuit, during the interval to point 5, is charged at the corresponding rate, mentioned above. From point 5 to point 6, the operations are repeated as they were in interval 2 to 3 (from point 2 to point 3). At point 6, the remainder has changed to $H = 1$, and on interval 6 to 7 the rate of charge of the integrating circuit has decreased by a factor of 2 as compared to interval 3 to 4. At interval 7 to 8, the charge of the integrating circuit is again in accordance with code $R = 3$, and then the process repeats.

As shown in Figure 3.9, if to take samples of the function E at some equal intervals of time T_c in the range of function values from U_{min} to U_{max}, where it is strictly linear, the values E_c of samples appear to be unchanged. U_{min} is the function value corresponding to point 4 (i.e., when the residual and the charge accumulated in the integrating circuit during this residual is minimal). U_{max} is the value of the function corresponding to point 8, that is, when its maximum value in the interval from the accumulator overflow to its next overflow is maximum among all possible cases for the chosen parameters R and Q.

The period of the mentioned samples is equal to $T_c = QT_r/R$ (i.e., their frequency is equal). Because of this, if the signal from the output of the integrating circuit is fed to the analog input of the sampling-stored pulse PD and its other input is connected to the pulse output of the VCO included in the PLL (it is when output 1 in Figure 3.9 is used), its frequency F_c will be brought, by the control voltage E_c, into accordance with the reference frequency through the frequency ratio obtained above. The frequency step size is equal to $dF = F_r/Q$. The working area of the static characteristic of the PD, extending from U_{min} to U_{max}, is wide enough for successful operation of the PLL system.

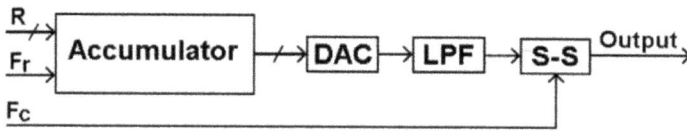

Figure 3.10 Variant with pulse PD of the sampling-storage type.

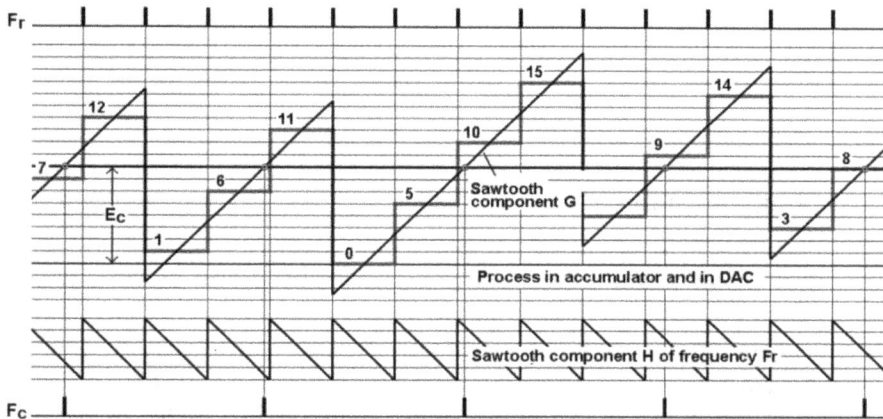

Figure 3.11 Timing diagrams explaining the detector's work in Figure 3.6.

It is clear that, in the case of a real integrator, due to its imperfection, a distortion of the process E at its output arises, which leads to fractional noise. The magnitude of this interference will be discussed below.

In principle, the frequency inputs F_r and F_c in Figure 3.9 can be swapped over in order to get a higher frequency at the PLL. However, it should be kept in mind that, in this case, the integrator appears to be included in the loop and, having a signal delay, may worsen the stability of the system.

If in the considered scheme the pulse-phase detector of sampling-storage type is used, then there are no objective reasons for including a narrowband lowpass filter in the PLL system. It can be used only for suppression of components with frequencies F_c and F_r, leaking

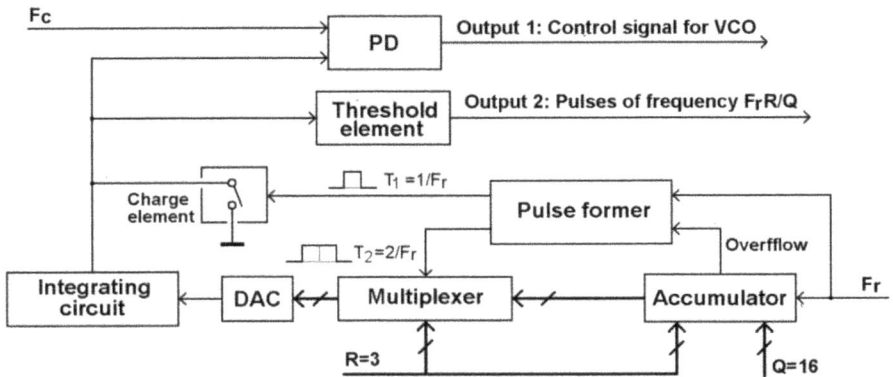

Figure 3.12 Nikiforov's scheme.

through the PD, but these frequencies are high enough, the lowpass filter can be broadband, and the PLL has high speed.

In another possible application of the described circuit, it can be used as a direct frequency synthesizer of pulsed signals. For this purpose, the signal from the output of the integrating circuit goes to output 2 (in Figure 3.9) through a threshold element. In this case, the pulse frequency at the scheme output is equal to $F_c = RF_r/Q$ and the frequency step size is equal to $dF = F_r/Q$.

Capacitance Q of the accumulator and number R on its input can both be variable, which extends the possibilities in obtaining a fine frequency resolution.

One more important and positive feature of the considered scheme is that the DAC operates in a much narrower range of code values than in many other systems, including, for example, DDS. Although the process in the accumulator occurs within its capacity Q, but the full scale of voltages at the DAC output corresponds to the number R, which is much smaller than the number Q. This reduces the accuracy requirements of the DAC or correspondingly increases its efficiency in suppressing fractional noise.

In the mid-1980s, the scheme was implemented in a series-produced product of the former USSR. Then the domestic element base was used. At the clock frequency $F_r = 625$ kHz, the frequency ranged from 118 to 156 kHz with a step size of 10 Hz, and fractional interference level at the output not worse than -70 dBc was provided. In

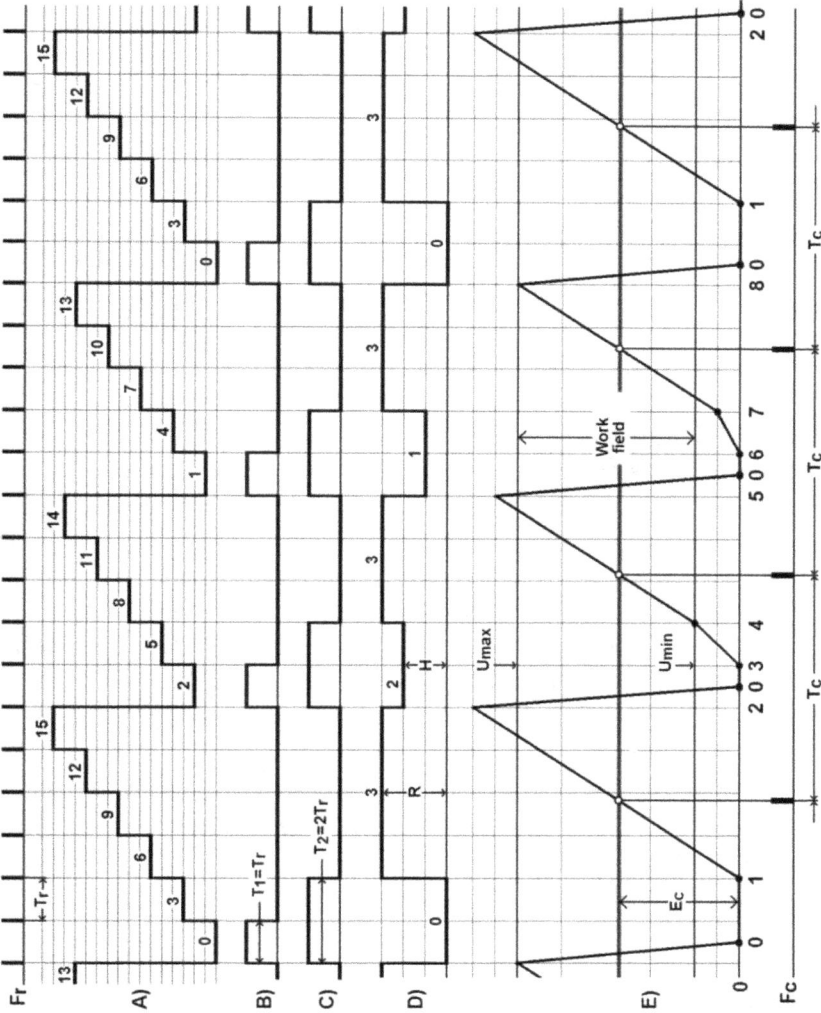

Figure 3.13 Diagrams explaining the operation of the scheme in Figure 3.8.

principle, the frequency resolution could have been as small as desired for the same level of interference, but such a task was not set. If to extrapolate this result to the current level of the world-integrated technology, we can expect quite good prospects for the use of the described scheme, up to its ability to compete, in some applications, with synthesizers such as DDS.

3.7 KOSLOV'S SCHEME

The scheme of the synthesizer is presented in Figure 3.14 [10–12], and Figure 3.15 shows the diagrams for explanations on operation of the scheme.

A frequency divider with a fractional variable division factor is presented here as a frequency divider FD clocked by reference frequency F_r and an accumulator clocked by output pulses of FD. The coefficient of frequency division equal to $N = N_0 - n$, where N_0 and n are accordingly integer and fractional parts of the division factor. As an example, the value N is $N = 3 - 1/4$. It is important to notice that in the formula of N instead of plus n, as usual, here it is minus n.

The fraction in the division factor is generated by an accumulator. In order to get the value $n = 1/4$, the capacity Q of the accumulator is equal to $Q = 4$. With each overflow of accumulator the coefficient N decreases by a unit, and this results in the fractional value of N.

The calculator of a numerator A of the fraction N is an arithmetic device. Its clock frequency is not important, in principle, and is therefore not shown in the figure. Having values N_0, n, and Q at its inputs, this block calculates the value of A. In the example, $N = 11/4$ (i.e., $A = 11$). The numerator A of this fraction is proportional to the reference frequency F_r, and the denominator is proportional to the signal frequency F_c formed in PLL.

The diagrams of Figure 3.15 show how the signal E_c is formed for controlling the frequency of the VCO in PLL. The digital signal $n(t)$ that comes from the accumulator is delayed by one cycle of F_r in a shift register, and then an adder adds number A to its value, thereby forming the signal $n(t + T_r) + A$. Next, a multiplexer, switched by the RS flip-flop, alternates the values $n(t)$ and $n(t + T_r) + A$. The RS-flip-flop is clocked by pulses of the signal frequency F_c and pulses from the F_r' output of FD, which has equal frequency to F_c, on the average.

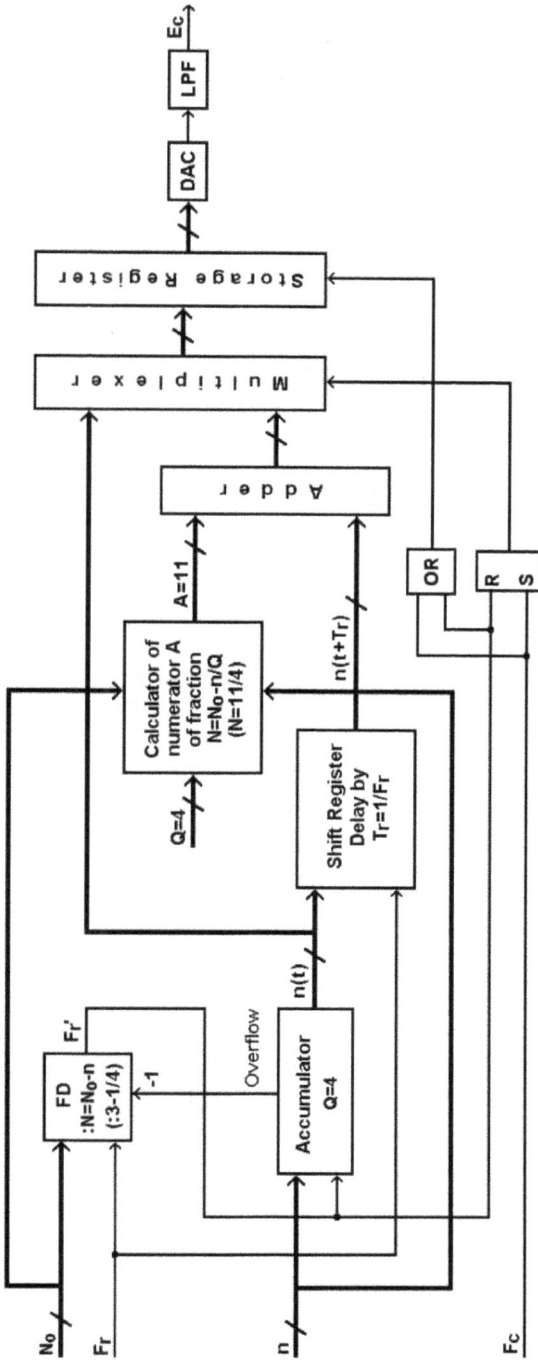

Figure 3.14 Koslov's scheme.

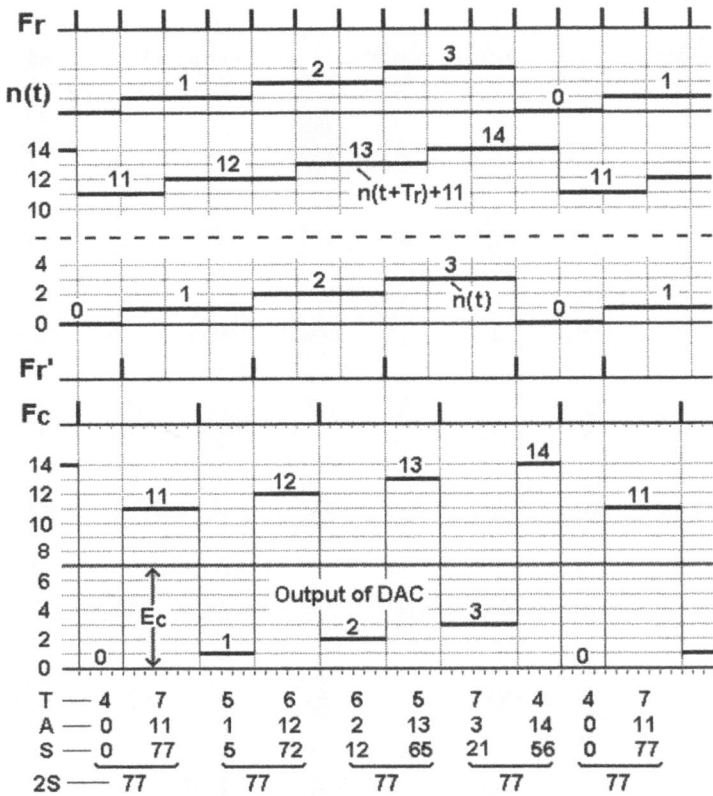

Figure 3.15 Diagrams for explanations on how the scheme works.

The signal of the multiplexer is recorded in a storage register clocked by the sum of pulses F_c and F_r' produced in an OR gate. From the output of storage register the digital signal comes to DAC, where it is converted to analog equivalent. A lowpass filter at the DAC output separates constant component E_c which is used for controlling the frequency F_c of the VCO included in the PLL.

The table at the bottom of Figure 3.15 is shown to explain the fractional noise suppression mechanism. It indicates, in conventional units: T (the width of pulses of larger and smaller levels), A (the height of pulses of larger and smaller levels), S (square of pulses of larger and smaller levels), and $2S$ (the sum of the squares of the two adjacent pulses of larger and smaller levels). The table shows that the square $2S$ in each signal period T_c remains constant, which is a prerequisite for effective suppression of the fractional components at the scheme output. Strictly speaking, the process at the DAC output is not exactly periodic. However, with the increasing length of the sequence, it becomes close to periodic with the $1/F_c$ period.

Although, in Figure 3.14, in order to simplify the explanation of the principle of circuit operation, the capacity of the accumulator in divider N is limited to only 2 bits, but, in reality, it can be much greater if it is necessary to provide the desired frequency resolution. If the accumulator is, for example, of 32 bits, then the step size is defined by changing N by 1/232 of its value. At the same time, there is no need to use all 32 bits from the output of the accumulator for the suppression of the fractional noise. It can be only used (the more significant bits) in an amount equal to the number of the DAC bits.

Table 3.2 presents calculations of the fractional noise spectrum at the output of the scheme, shown in Figure 3.14, normalized relative to the capacity of the DAC. This is done for coefficients N equal to 3 − 1/16, 3 − 1/32, and 3 − 1/64. The number q of DAC bits is equal to, respectively, 6, 7, and 8. Selected ratios are characterized by the fact that the code at the input of the accumulator looks like 00... 01, and so the first harmonic of fractional interference is of the highest level relative to other ones in the bandwidth of PLL, and thus it is possible to judge the efficiency of the scheme action. Table 3.1 shows calculations for the ideal DAC, with no errors, and for the "real" one having in the MSB an error equal to value of the LSB (−1 LSB).

Calculation of the spectrum when the DAC is of more than 8 bits is difficult. However, as can be seen from Table 3.2, there is a simple trend of decreasing the level of first harmonic by 6 dB with each increase in the number of DAC bits by one. So, as an example, Table 3.2 shows the possible interference level when using a 12-bit DAC.

In fact, the scheme of Figure 3.14 is a PD having the static characteristic slope of $K = U/2\pi$, where U is the full scale of DAC output voltage. Being included in PLL, the scheme produces the spurs distributed as two sidebands with levels larger by $20\lg\pi = 10$ dB than shown in Table 3.2.

As mentioned above, Table 3.2 shows the worst case of code n when it has the form $n = 00...01$. In the other possible cases, the interference level may be much smaller. For example, Table 3.3 shows the level of fractional interferences when the code n is $n = 010...01$. In this case, the fractional noise decreases by an order of magnitude.

Table 3.2

Harmonic number		0	1	2	3	4	5	6	7	8	9
q=6	DAC of ideal	-6.0 (E_c)	-58.2 (1/16Fc)	-78.4	-60.7	-55.7	-52.6	-50.1	-47.9	-45.8	-43.8
	DAC of real	-6.0	-50.5	-62.3	-69.6	-58.5	-54.0	-51.0	-48.5	-46.3	-44.1
q=7	DAC of ideal	-6.0 (E_c)	-62.7 (1/32Fc)	-72.2	-74.9	-70.2	-66.8	-64.4	-62.5	-60.9	-59.5
	DAC of real	-6.0	-56.2	-64.1	-71.5	-75.3	-70.4	-66.7	-64.1	-62.1	-60.4
q=8	DAC of ideal	-6.0 (E_c)	-68.3 (1/64Fc)	-75.8	-82.5	-91.0	-87.5	-82.0	-78.5	-76.0	-74.0
	DAC of real	-6.0	-62.0	-68.7	-73.6	-78.6	-85.9	-95.8	-84.5	-79.4	-76.3
- - -											
q=12	DAC of ideal	-6.0 (E_c)	-92.0 (1/1024Fc)	(Estimated, extrapolation)							
	DAC of real	-6.0	-86.0								

Level, dB

Table 3.3

	Harmonic number	0	1	2	3	4	5	6	7
	q=7 n=00001 DAC of ideal	-6.0 (Ec)	-62.7 (1/32Fc)	-72.2	-74.9	-70.2	-66.8	-64.4	-62.5
	DAC of real	-6.0	-56.2	-64.1	-71.5	-75.3	-70.4	-66.7	-64.1
	q=7 n=01001 DAC of ideal	-6.0 (Ec)	-88.6 (1/32Fc)	-85.7	-79.8	-72.0	-56.2	-65.3	-65.1
Level, dB	DAC of real	-6.0	-78.6	-86.4	-102.1	-122.8	-58.3	-67.3	-66.6
	q=12 n=0100000001 DAC of ideal	-6.0 (Ec)	-112	(Estimated, extrapolation)					
	DAC of real	-6.0	-102						

This leads to the idea of using the preferable codes. If in one of the paths, the reference or signal one, to add a frequency divider with a division factor N_1, then it is possible to get substantially the same signal frequency F_c with various combinations of the coefficients N_1 and N. Among these combinations, there can be chosen the most appropriate ones, in terms of the minimum level of fractional noise.

Figure 3.16 shows an example of the functional diagram when an additional divider of coefficient N_1 is included in the signal path of the synthesizer. Basically, in its structure, it may be similar to the fractional divider of ratio N (Figure 3.14), but it is not necessary to have in it a large capacity of accumulator; it is possible to limit it, for example, by only 5 bits. Then the fractional interferences generated by this divider are sufficiently of high frequency to be successfully filtered by PLL. To facilitate the suppression of the interferences, there can be also use of one of the suppression schemes, such as Cox's scheme.

Let's suppose, for example, that, for obtaining some signal frequency, it is required to have a ratio $N_s = N_1/N = 0.888$, and the control code n in the accumulator of the frequency divider of ratio N is of the preferable type as $n = 010...01$, that follows from Table 3.3. It is quite enough that this kind of the code is applied only to the five more significant bits of the accumulator, because just they determine the character of the signal spectrum. Then this 5-bit code n has to be equal to $n = 01001$. The numerical value of this code, relative to a capacity of 5 bits, is 9/32, and, thus, the value of the coefficient N is equal to $N = 3 - 9/32 = 87/32$. Using the given value of the ratio $N_s = 0.888$, the coefficient N_1 can be calculated as $N_1 = 2.414$. This means that the numerical value of code $n1$ in the accumulator of the divider of ratio $N1$ should be equal to $1 - 0.414 = 0.586$, which is approximately number 19 in the 5 bits of the accumulator. In turn, the resulting number corresponds to the code $n_1 = 10011$ of this accumulator.

Using this algorithm of calculations, we can obtain a preferred value of code $n1$ for any relationship N_s. The obtained values of code $n1$ can be introduced into a table placed in the interface of the synthesizer and can be used when setting the frequency of the signal. It should be clear that, in order to hold the value of the code n, equal to $n = 01001$, as shown in the above example, the table must include 32 values of code $n1$, each of which belongs to one of 32 subbands of the signal frequency.

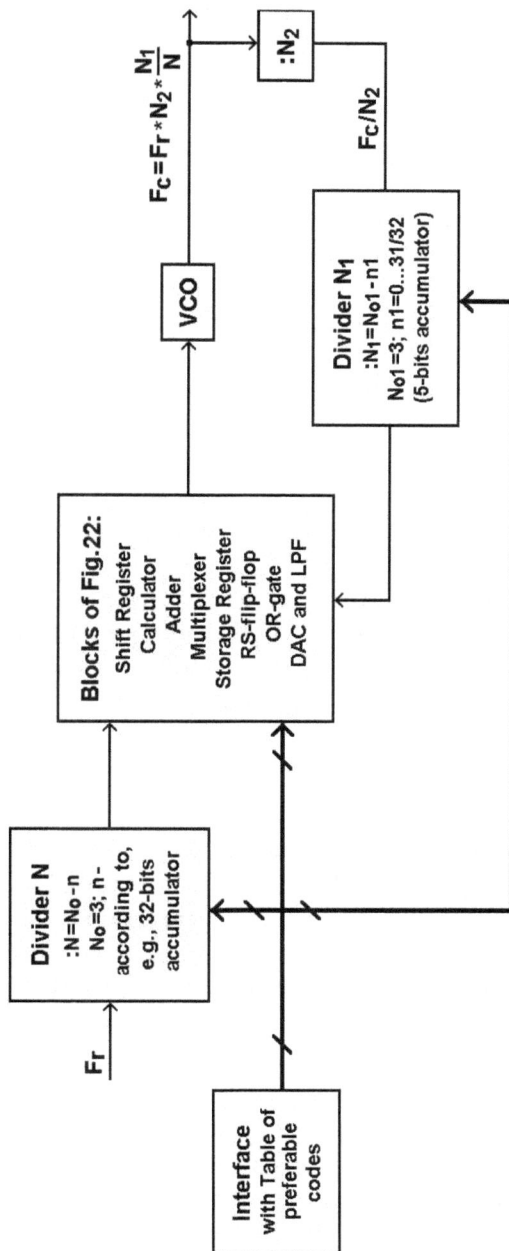

Figure 3.16 Functional diagram of the synthesizer using the preferable codes.

In a real synthesizer, the reference frequency can be chosen, for example, as $F_r = 1.6$ GHz. Then the operating frequency of the DAC will be in the range from about 533 to 800 MHz. To get, for example, the signal frequency by order of 10 GHz, it is necessary to include in PLL a prescaler with a coefficient N_2, and the total division factor in the signal path has to be equal to about 12. As a result, taking into account the data on the level of interferences given in Table 3.3 for the real 12-bit DAC, the spurs level at the output of the synthesizer are not to exceed $- 102 + 10$ (account of PD slope) $+ 22$ (multiplication by 12 times in the loop) $= -70$ dBc, which can be considered a quite good result for such a simple structure of the single-loop synthesizer.

This variant of suppression of the fractional noise has the distinct advantage before well-known schemes because there are no analog circuits in the scheme for suppression that require high accuracy of their interconnection. The whole process here is running in a digital form having, in principle, no errors, and all depends only on the accuracy of the DAC.

It is natural to refer to the analogy of the discussed structure, which may be named, for brevity, as Frac-N-Syn, with a synthesizer based on PLL with DDS included in the loop as a fractional frequency divider. First, DDS with such small coefficients of division, such as $N = 3$, is, it may be said, unworkable at all because it provides fractional noise hardly better than -30 dBc and even by 10 dB worse, if bringing the signal frequency, say, up to even if about the clock frequency. Second, the use of large values of the coefficient N in the DDS leads to the same large multiplication factor if the signal frequency needs to be raised to the order of the reference one. Respectively, it leads to the respective multiplication of noises at the output of the PD that naturally degrades the signal-to-noise ratio. Third, the DAC in Frac-N-Syn operates at a lower frequency, by N times less than the reference one; therefore, it has increased its accuracy and hence reduced fractional noise. Fourth, Frac-N-Syn is much simpler than the DDS.

If comparing the Frac-N-Syn with the fractional-N PLL structure, then it has the following advantages: (1) the absence of DSM, with which therefore there is no need to reduce the PLL bandwidth for its filtering and this provides more fast frequency switching; and (2) the use of small division factors in the loop, which significantly reduces noise in the signal spectrum.

In conclusion, Frac-N-Syn has a good prospect of being embodied in an integrated circuit or field-programmable gate array (FPGA) to successfully replace the DDS in some areas of applications.

3.8 SIMPLIFIED VERSION OF THE DDS

As mentioned above, the synthesizer of the DDS type can be used in complex structures to generate a relatively small frequency range with high-frequency resolution.

The DDS possess a unique set of such positive properties as digital control of the frequency and phase of the synthesized signal, high frequency and phase resolution, and extremely fast switching to a new frequency and phase without phase disruption. Thanks to these advantages, the world market is constantly replenished with new developments of DDS chips with continuous improvement of their characteristics as integrated technology develops. Among the many companies, the leader in this area is Analog Devices, Inc. Knowing about the main characteristics of existing DDS chips, it is enough to see this company's datasheets.

The functional core of a conventional DDS consists of a phase accumulator, a sinusoidal table (lookup table (LUT)), and a DAC. A suggested simplification of the DDS structure is to exclude the LUT from it. There, a step-sinusoidal function is not formed, but a step-triangular one is and is further converted, as in a regular DDS, into a step-by-step analog signal, smoothed at the output by a lowpass filter [13, 14].

The scheme of such a variant of DDS is shown in Figure 3.17, and its operation is explained with the time diagrams shown in Figure 3.18.

The accumulator with each pulse of the reference frequency F_r forms a digital sawtooth function represented by diagram A. In the simplified example, the accumulator is of 3 binary digits ($q = 3$), that is, its capacity Q is equal to $Q = 2^q = 8$ bits. The value of the steps of the generated function is equal to the value of the control code R at the input of the accumulator, and the code, in this example, is equal

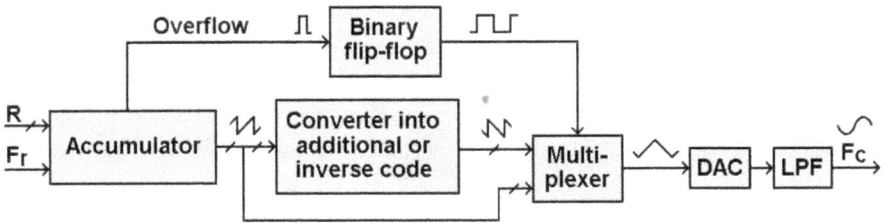

Figure 3.17 Simplified version of DDS.

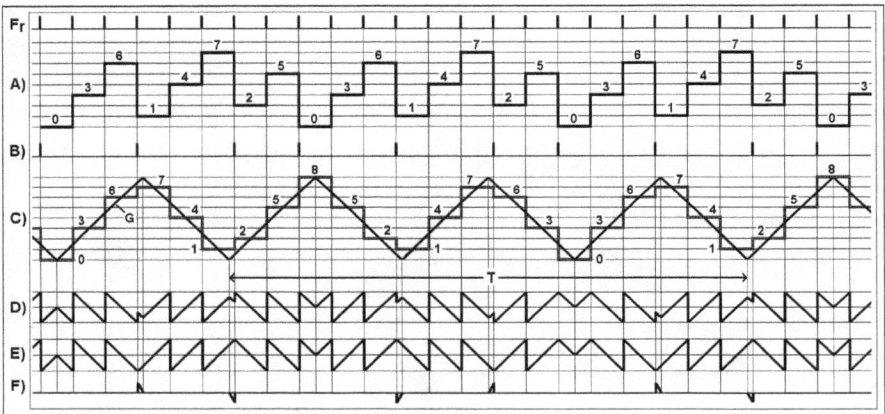

Figure 3.18 Timing diagrams explaining the operation of the circuit in Figure 3.17.

to $R = 3$. From the output of the accumulator, the step-by-step process passes to the converter, which transforms it into an additional or inverse code, and from it the converted code comes to one of the inputs of a multibit multiplexer. The other input of the multiplexer is connected to the output of the accumulator. The multiplexer alternates the digital processes arriving at it with the help of a binary flip-flop, which is triggered by overflow pulses from the accumulator (diagram B). In this way, a digital triangular function is formed, which then, using the DAC, turns into its analog equivalent.

The resulting process at the output of the multiplexer, for the case of an additional code, as well as its corresponding analog at the output of the DAC, is shown in diagram C. It is the sum of two components: a periodic triangular function G with a full amplitude equal to Q, and a period equal to two cycles of filling the accumulator, as well as a complex process (diagram D) with a full amplitude equal to R. The latter contains a phase-modulated sawtooth function (diagram E) and a low-frequency pulse signal (diagram F). Process E mainly includes high-frequency components eliminated by the lowpass filter, but the spectrum of low-frequency pulses of the diagram F lies completely in the lowpass filter band and is the main reason for the deterioration of the signal spectrum at the output of DDS. The full period T of the process C is $2Q = 16$ clocks of F_r, and $R = 3$ periods of the triangular function G are placed on this interval, which, being the third harmonic of the process C then after the lowpass filter is transformed into a sinusoidal signal with a frequency $F_c=F_rR/(2Q) = 3F_r/16$.

When the levels of the higher harmonics of the signal are small enough, they decrease in proportion to the square of the harmonic number. Therefore, the lowpass filter can be simple and fairly broadband, for example, with an octave overlap. The greater the $2Q/R$ ratio, that is, the lower the signal frequency relative to the clock frequency, the more accurate, due to the greater number of steps, the triangular function is approximated, and the higher spectral purity of the signal.

Figure 3.19 shows the signal spectrum for the case of a 7-bit accumulator ($q = 7$) and the control code $R = 13$, when converting the accumulator output to an additional code. The calculations are carried out in the FFT program. The calculations were made both with an ideal DAC that does not have errors and with a real one, hereinafter referred to as a converter with an error in the MSB equal to -1 of the LSB. The working signal of the synthesizer is at the 13th harmonic, and its number coincides with the numerical value of the code R. With an ideal DAC, even harmonics in the spectrum are absent in principle. Figure 3.19 also does not show harmonics with negligible levels below -80 dBc. With a real DAC, even harmonics appear; however, their levels are negligible, below -70 dBc. In general, spectrum degradation with real DACs is negligible. In the worst case, at the ninth harmonic, the excess relative to the case of an ideal DAC is 1.5 dB. The third harmonic of the signal, the 39th in the spectrum, has a value of -19 dBc, as it should be in accordance with FFT for the triangular function.

The harmonic spectrum is also calculated for the same initial data, q and R, when the accumulator output is converted to an inverse code. The results of calculations with a real DAC are summarized in

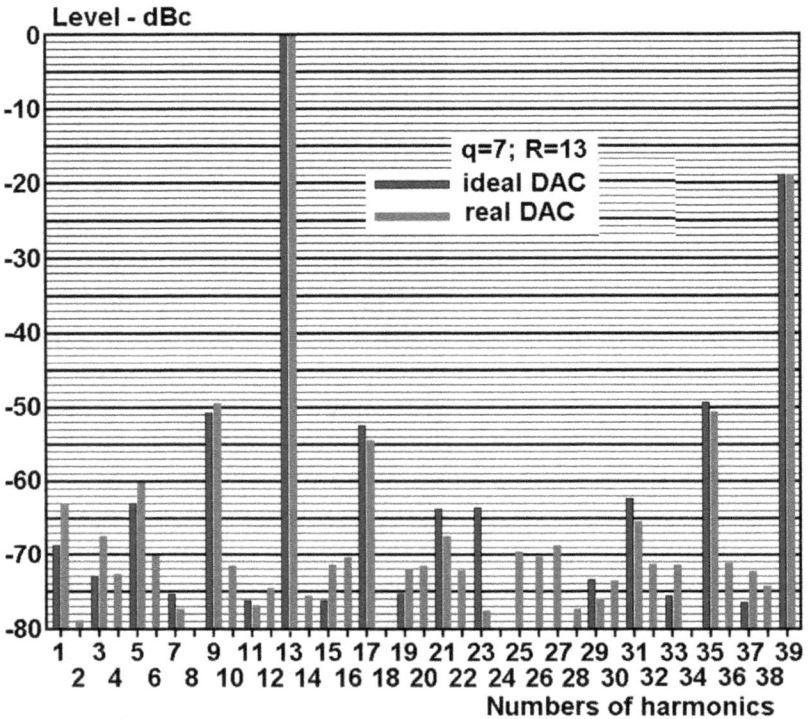

Figure 3.19 Signal spectrum.

Table 3.4, which shows a comparison of spur levels with additional and inverse codes. Even harmonics, because of their negligible level, are not included in Table 3.4.

Table 3.4 shows that the difference is insignificant. For the worst case, at the ninth harmonic, the difference is only 0.1 dB in favor of the inverse code. It is also important to note that the operation of obtaining an inverse code is much simpler than generating additional

Table 3.4

Code	Harmonic Number and Level (dBc)									
	1	3	5	7	9	11	13	15	17	19
Additional	–63.5	–67.5	–60.1	–77.7	–49.5	–76.9	0	–71.5	–54.4	–72.0
Inverse	–63.9	–71.2	–59.9	–77.2	–49.6	–70.5	0	–71.9	–54.8	–72.1

code. In addition, the DAC is also simplified, since there is no need for an additional MSB; it is for the 2^q digit. This implies that, in practice, it is more preferable to use an inverse code, and we will deal only with this kind of code.

The above cases were considered when the digital process from the output of the accumulator, all its bits, were converted to analog using the DAC. However, due to the limited capacitance of the DAC, and, at the same time, of the accumulator, it is impossible to provide the desired high-frequency resolution of the synthesized signal. The problem is solved by using only the more significant bits of the accumulator for code conversion with its full unlimited and sufficiently large capacity. Then the carry of the overflow unit from less significant bits to MSBs affects the spectrum in the same way as the inaccuracy of the DAC as shown above, and thus the degradation of the spectrum is insignificant. This can be shown with a simple example.

Table 3.5 shows the signal spectrum at $R = 3$, $q = 5$ and the corresponding of 5-bit DAC, and Table 3.6 shows the spectrum when 2 LSB bits, not having an output to the DAC, are added through the carry circuit, to the existing 3 MSB bits of the accumulator. The control code in this case is equal to $R = 3 + 1/4$, that is, the value of code at the input of the less significant bits is equal to 1.

In Table 3.5, the signal is the third harmonic of the process lasting 64 periods of the reference frequency, its number coincides with the value of the control code $R = 3$. In Table 3.6, its harmonic number turns out to be 13 in accordance with the code value $R = 13/4$ when a process duration is of 256 periods. On the frequency axis, the signals are nearby; the 13th harmonic of Table 3.6 is slightly higher in frequency of the third harmonic of Table 3.5.

From the consideration of the previously mentioned tables, the conclusion follows that the spectrum deteriorates insignificantly when an overflow unit transfers from the LSB block to the MSB

Table 3.5

Harmonic Number and Level (dBc)									
1	2	3	4	5	6	7	8	9	10
−49.6	−51.9	0	−51.9	−48.4	−52.0	−47.1	−52.1	−17.9	−52.2

Table 3.6

Harmonic Number and Level (dBc)									
1	2	3	4	5	6	7	8	9	10
−56.0	−51.6	−51.0	−56.1	−51.4	−64.8	−48.7	−47.4	−48.4	−61.8
11	12	13	14	15	16	17	18	19	20
−54.2	−62.7	0	−53.3	−63.0	−53.6	−48.3	−61.7	−47.0	−61.9

block. Therefore, there is no problem of achieving an arbitrarily high-frequency resolution by increasing the number of LSBs that do not have an output to the DAC. As a result, we can obtain formulas for the frequency F_c of the signal and the frequency resolution dF:

$$S_r = U_{max} R / (S \pi Q_r) \text{ and } S_c = U_{max} C / (2 \pi Q_c)$$

where Q is the total capacity of the accumulator and R is the integer code value relative to the capacity Q.

The calculations of the spectrum for DAC of higher orders, above 7 bits, and therefore the MSBs of the accumulator of the same capacity, are difficult because of their complexity. However, when calculating the spectrum at their relatively small orders, less than 8 bits, a simple regularity appears: the spectrum improves by 6 dB with each octave decrease in the $R/2Q$ ratio. The last ratio is actually the ratio of the resulting operating frequency F_c of the synthesizer to its clock reference frequency F_r, that is, F_c/F_r.

Following the regularity noted above and the initial data on the worst case of the calculated spur level −47 dBc with a block of 5 bits of the real DAC and the same number of bits of the MSB block of the accumulator (see Table 3.6) when $F_r/F_c = {\sim}20$, it is easy to calculate that, for example, at 6 bits of these blocks and the ratio $F_r/F_c = 40$, it is possible to guarantee a level of spurs not more than −53 dBc. It is possible to get a lower level of spurs, increasing the capacity of the named blocks. At 8 bits, for example, it is possible to achieve a decrease in the level to about −65 dBc. True, in this case, the signal frequencies decrease by 4 times, respectively.

The calculated results cannot be considered as outstanding, but it is important to note how easy they are achieved with low enough requirements for the bit capacity and accuracy of the DAC. Comparing these results with the capabilities of conventional DDS is difficult. The fact is that manufacturers of DDS chips do not give guarantees for the maximum level of spurs. Only spectral diagrams at certain frequencies are given without explanation of the selection criteria for these frequencies and do not show the worst cases, that is, with the highest level of spurs. True, Analog Devices Inc., for example, can give such guarantees, but they are subject to an additional fee for testing each chip of a purchased batch. The money, perhaps, is considerable, especially with a large batch of ordered chips.

Analog Devices Inc. also has an ADIsimDDS program for spectrum calculation at any signal frequency. It does not take into account all the possible inaccuracies of the conversion, but can give an approximate idea of the capabilities of the classic DDS. As an example, take one of the latest ADI developments, the AD9914 synthesizer with an integrated 12-bit DAC, operating at a clock frequency of 3.5 GHz with a consumption of about 3W. If we are following the program ADsimDDS, it has an advantage over simplified DDS at relatively high signal frequencies. Until the ratio $F_r/F_c = 40$, the spurs are approximately −53 dBc, and further, with an increase in this ratio, it remains the same. In the simplified version, the spurs decrease by 6 dB with each octave increase in the F_r/F_c ratio. Thus, the ratio $F_r/F_c = 40$ may be the boundary of the transition of the spectral purity advantage from traditional DDS to its simplified version.

In the simplified version of DDS considered here, the same as in the conventional one, the frequency and phase modulations with high speed are easily implemented. For phase modulation, a digital adder can be enabled between the accumulator and the code converter controlled by the phase code. Also it can be done without such an adder, if acting on the phase of the signal in a known manner by short-term positive and negative increments to the frequency of the signal.

In addition to the main synthesizing functions for obtaining a spectrally pure sinusoidal signal with the ability to quickly switch frequencies with high-frequency resolution, the considered option, such as the conventional DDS, can be used to generate signals and a nonsinusoidal form. The meander can be obtained from the output of the binary flip-flop, and short pulses can be obtained from the output of the accumulator overflow. Sawtooth and triangular signals are realized at the main output with a corresponding expansion of the band of the lowpass filter up to the clock frequency. In the case of a saw-tooth signal, the multiplexer is switched to the external control mode (the binary flip-flop is disconnected from its control input), which makes it possible to obtain a falling or rising form of the signal. Also, such a device can be used as a frequency divider in PLL-based frequency synthesizers.

The considered version of the DDS performs the basic functions of synthesizers of this type, such as fast switching of frequency and modulation in frequency and phase, and provides high-frequency resolution and a satisfactory level of spectral purity for possible applications. Moreover, its structure is much simpler and more economical in terms of power consumption. It can be made both as an integrated chip or on an FPGA with an external standard DAC.

References

[1] Braymer, N. B., "Frequency Synthesizer," US Patent 3,555,446, January 12, 1971, filed January 17, 1969.

[2] Gillette, G. C., "Frequency Synthesizer System," US Patent 3,582,810, January 6, 1971, filed May 5, 1969.

[3] Zhuk, O. Y., and V. I. Koslov, "Digital Frequency Synthesizer," Patent of Russia No. 470901, priority 12.01.73, in Russian.

[4] Cox, R. G., "Frequency Synthesizer," US Patent 3,976,945, August 24, 1976, filed September 5, 1975.

[5] Underwood, M. J., "Frequency Synthesizer," UK Patent 1447418, August 25, 1976.

[6] Koslov, V. I., "About One Method of Frequency Synthesis," *Electrosvyaz*, No. 112022, in Russian.

[7] Koslov, V. I., "Digital Phase Detector," Patent of Russia No. 894854, priority 06.02.80, in Russian.

[8] Nikiforov, V. I., "Frequency Synthesizer," Patent of Russia No. 1415410, priority 11.01.85, in Russian.

[9] Nikiforov, V. I., "Frequency Synthesizer," Patent of Russia No. 1501265, priority 10.11.87, in Russian.

[10] Koslov, V. I., "Method for Phase Detection of Pulse Sequences at Unequal Frequencies and a Device for Its Implementation," Patent of Russia No. 879738, priority 21.05.79, in Russian.

[11] Koslov, V., "A Fractional-N PLL Synthesizer without Delta-Sigma Modulation as a New Concept in Frequency Synthesis," *Microwave Product Digest*, April 2015.

[12] Koslov, V. I., "Frequency Synthesis with Digital-to-Analog Suppression of Fractional Interference in the PLL System," *Electrosvyaz*, No. 6, 2020, in Russian.

[13] Koslov, V. I., "Frequency Divider-Synthesizer," Patent of Russia No. 1149395, published 12/08/84, priority 11.10.82, in Russian.

[14] Koslov, V. I., "Simplified Version of the Direct Digital Frequency Synthesizer," *Electrosvyaz*, No. 7, 2020, in Russian.

4

THE IDEA OF A MULTIFREQUENCY
PHASE DETECTOR

4.1   IDEA BY BOSSELAERS

Bosselaers is, perhaps, the author of the first patent in which an attempt was made to approach the creation of a multifrequency phase detector (MFPD) capable of operating at unequal comparison frequencies. The scheme for his patent is shown in Figure 4.1 [1]. It uses, as a phase detector in the PLL, two accumulators, one each in the reference and signal paths, and a subtractor of the current contents of these accumulators.

Accumulator R is clocked by the reference frequency F_r, and accumulator C is clocked by the frequency of the signal F_c from the PLL VCO. The inputs of the accumulators receive the codes R and C, respectively, of which each numerical value is added to the current content of the corresponding accumulator with each clock of the corresponding frequency. The accumulator outputs are connected to a subtractor clocked by the reference and signal frequencies. In it, under the action of clocking by signal pulses, the current content of the accumulator C is added to the current state of the subtractor, and under the action of clocking by the pulses of the reference frequency, the current content of the accumulator R is subtracted from it.

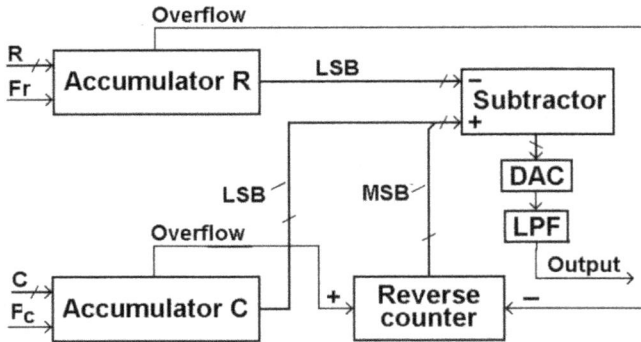

Figure 4.1   The structure of the PD according to Bosselaers's patent.

There is also a reverse counter, the purpose of which is as follows. Each of its two inputs is connected to the overflow output of the corresponding accumulator. When the accumulator C overflows, the state of the counter increases by one, and when the accumulator R overflows, it decreases by one. Thus, the difference between the two most significant bits of the processes in the subtractor is formed at the output of the counter, which, accordingly, makes the most significant bits of its output process. The output of the subtractor goes further to the DAC to obtain a voltage that controls the frequency of the VCO.

The principle of operation of the detector, shown in Figure 4.1, can be explained using the timing diagrams shown in Figure 4.2. To make it easier to understand, a simple example with small values of the codes $R = 5$ and $C = 4$ at the 4-bit accumulator inputs is chosen. Figure 4.2 shows the process at the output of the subtractor or the voltage proportional to it at the output of the DAC.

Having done simple geometric constructions, it is easy to make sure that, at the DAC output, in addition to the constant component E_c used to control the VCO frequency, there are only two sawtooth components with the original frequencies F_r and F_c, which, due to their rather high frequencies, can be easily eliminated by a lowpass filter before reaching the control input of the VCO.

The selected values of the codes provide the state of synchronism in the PLL system at the appropriate ratio of frequencies at the detector inputs, namely: $F_r/F_c = 4/5$. It is quite clear that if the ratio F_r/F_c turns out to be not equal to 4/5, then the PLL will go into tuning mode, bringing the VCO frequency to the specified ratio.

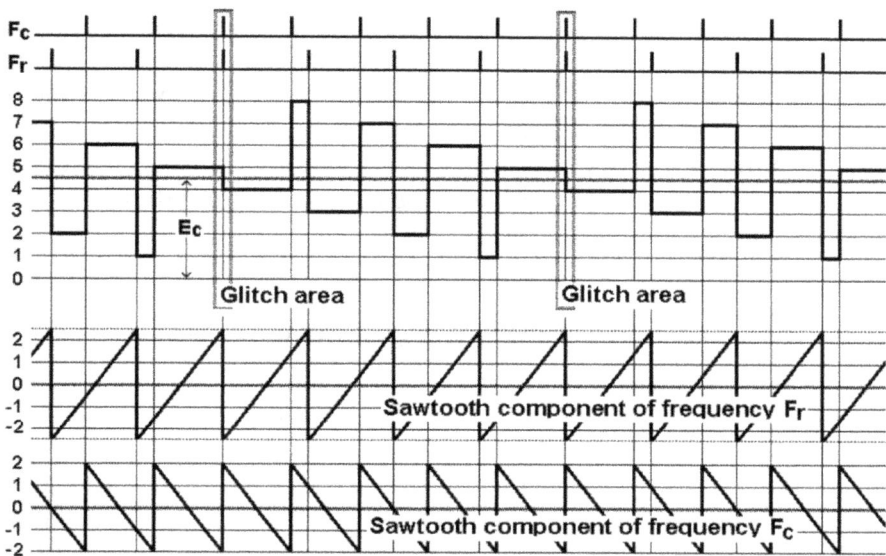

Figure 4.2 Timing diagrams explaining the detector operation.

There are no explicit frequency dividers in the proposed structure. However, their function is performed by accumulators, the result of which is their overflow impulses affecting the reversing counter. The latter plays the role of a phase detector itself, which switches under the influence of accumulator overflow pulses. The overflow frequencies of one and the other accumulators are equal to RF_r/Q and CF_c/Q, respectively, on average, where Q is the capacity of each accumulator. Therefore, frequency division is actually present, although it can be only a few units. The division ratios are, respectively, Q/R and Q/C, on average, and in the general case are fractional, since the number Q is most often not divisible by R and C without a remainder. In this case, the fractional interference is suppressed in the DAC, as follows from Figure 4.2.

It is clear that the quality of the spectrum of the synthesizer with such a detector depends, to a certain extent, on the accuracy of the DAC. However, in this scheme, this is not the main factor in its quality. The main disadvantage of the circuit is that at the moments of time when the positions of the pulses F_r and F_c coincide, it is impossible to unambiguously determine the result at the output of the subtractor. This is marked as the "glitch area" in the picture. At such moments of time, when a new current value of the total code from the accumulator C and the reverse counter arrives, the current value of the code from the accumulator R must be subtracted from it, which is practically impossible to do without an error. Because of this, a malfunction of the subtractor occurs, leading to a very significant deterioration in the spectral purity of the synthesized signal. Therefore, this structure has not been applied in practice.

4.2   IMPROVEMENT OF BOSSELAERS'S SCHEME

The scheme proposed by Bosselaers can be improved with the elimination of the noted drawback, as was done in [2, 3].

In the circuit in Figure 4.3, two identical accumulators are used: the reference one, accumulator R, and the signal one, accumulator C. In one of them, for example, in the accumulator R, the inverse value of the code is sent to the output, as shown in Figure 4.3. The peculiarity of the circuit is that the subsequent operations with the current values of the codes, taken from the outputs of the accumulators, are not performed in digital, as it is in the original version, in Bosselaers's circuit, but in analog form. For this, a summation-type DAC is used. It has, respectively, 2 inputs, so that in each of its bits the logical levels of the corresponding bits of the accumulators are summed, with the corresponding weight. Hereinafter, the DAC circuit is depicted, for simplicity and clarity, in the form of a resistive ladder in order to show the main thing, the weight values of the digits.

The most significant bit of the DAC is powered by an RS flip-flop, which is triggered by accumulator overflow pulses. The overflow

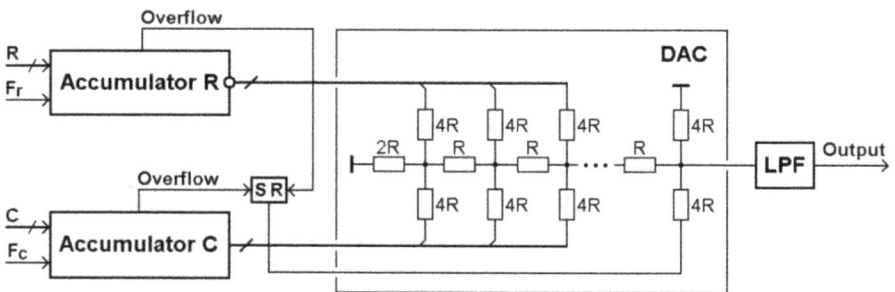

Figure 4.3 Improved scheme with analog summation.

pulse of accumulator C sets the flip-flop to 1, and the overflow pulse of accumulator R returns it to 0. The trigger, thus, acts as a reverse counter in Bosselaers's circuit. The constant component of the pulse process in the DAC is selected by the lowpass filter and fed to the PD output.

If we assume that each of the accumulator is 4 bits, and the values of the codes, as in the example with Bosselaers's circuit, are equal to $R = 5$ and $C = 4$, then at the output of the DAC there is a process that coincides with the one shown in Figure 4.2 to within a scale factor. However, failures in this circuit, due to the time coincidence of pulses F_r and F_c, are excluded, which makes the circuit quite suitable for practical use. Spectrum quality here depends only on the accuracy of the DAC.

4.3 VARIANT WITH ACCUMULATOR AND RS FLIP-FLOP

It is possible to further simplify the scheme as is shown in Figure 4.4 [4, 5]. The peculiarity of the circuit, in comparison with the previous version, is that only one reference accumulator is used here. The second signal accumulator is excluded due to the fact that the code C acts at its input, the value of which is chosen to be equal to the capacity of the accumulator C, and then the function of the accumulator is reduced to a simple transfer of signal pulses to the input of the RS flip-flop in that is there is no need for it. The diagrams in Figure 4.4 are shown for the case when the capacity of accumulator R is $Q = 8$, and the numerical value of the code at its input is $R = 3$. Figure 4.5 shows how it works.

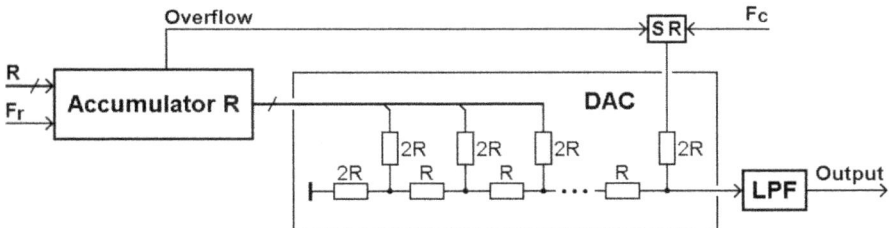

Figure 4.4 Variant with an accumulator and RS flip-flop.

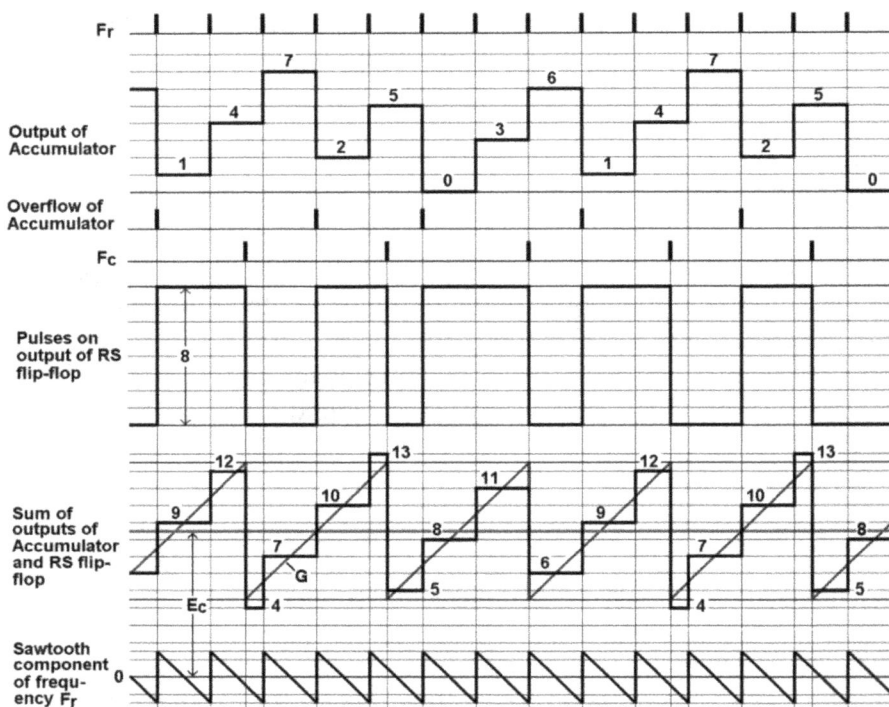

Figure 4.5 Timing diagrams explaining the operation of the circuit in Figure 4.4.

In the accumulator, under the action of the pulses of the reference frequency F_r, a step function is formed that is transmitted to the DAC. The accumulator overflow pulses are sent to one of the inputs of the RS flip-flop, setting it to the 1 state. The other input of the trigger receives pulses from the VCO with frequency F_c, which return the RS flip-flop to the 0 state. The pulses from the trigger output go to the most significant bit of the DAC, while the remaining bits of the DAC are fed by the pulses of the mentioned step function from the accumulator output. The resulting process in the DAC contains a constant component E_c and two high-frequency sawtooth components with frequencies F_r and F_c, which are eliminated, as in the previous versions, by a lowpass filter.

The advantage of this structure is the ability to use a conventional DAC; the characteristics of the spectral purity of the synthesizer signal will depend on it.

4.4 VARIANT WITH RING REGISTER

This version is based on the method [6, 7] illustrated by Figure 4.6.

Suppose that there is some logical structure that can perform the following operations:

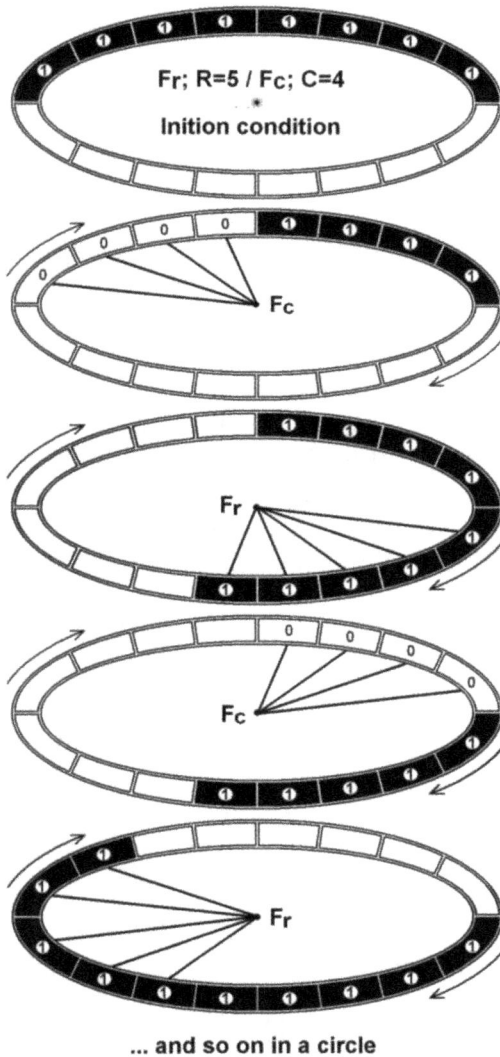

Figure 4.6 Explanation of the synthesis method.

- Accumulation of the number R of logical units at the arrival of each pulse of frequency, which, for example, can be considered as reference frequency F_r;
- Removal of the number C of the earliest from the accumulated logical units at the arrival of each pulse of frequency, which, for example, can be considered the signal frequency F_c.

Figure 4.6 shows it for the case of $R = 5$ and $C = 4$. In this case, the ratio of the frequencies F_r and F_c is, respectively, $F_r/F_c = 4/5$. As can be seen from Figure 4.6, $R = 5$ units is added to the "head" of the process at each clock of the frequency F_r and $C = 4$ units are removed at each clock of the frequency F_c from the "tail" of the process, and this happens in a circle.

The method can be used to construct an MFPD as shown in Figure 4.7, where the above operations are performed using a pulse distributor and a ring shift register. The current contents of the register are converted by the DAC to an analog equivalent to obtain the control voltage for the VCO in the frequency synthesizer circuit.

Figure 4.8 shows diagrams of the operation of such a structure. The diagrams correspond to the steady state in the PLL, and, in this case, the DAC output, as in the previous versions of the circuits, consist of only the DC component E_c, used to control the VCO frequency, and two sawtooth components with the frequencies F_r and F_c, suppressed by the lowpass filter.

It is important to note that, as follows from Figures 4.7 and 4.8, there is no frequency division in the PLL, the phase comparison takes place directly at the initial, in the general case, unequal frequencies. This is an important advantage of the circuit, since, being included

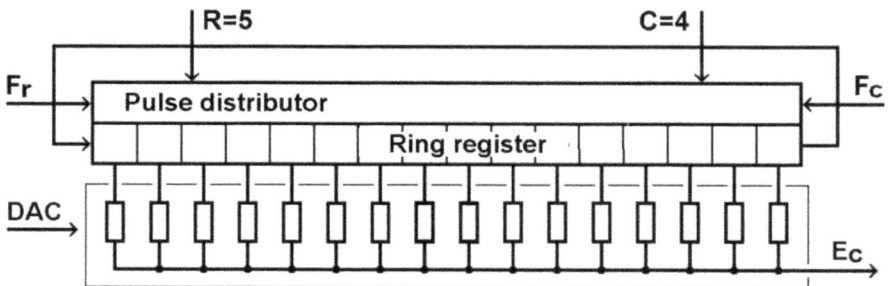

Figure 4.7 MFPD based on the ring register.

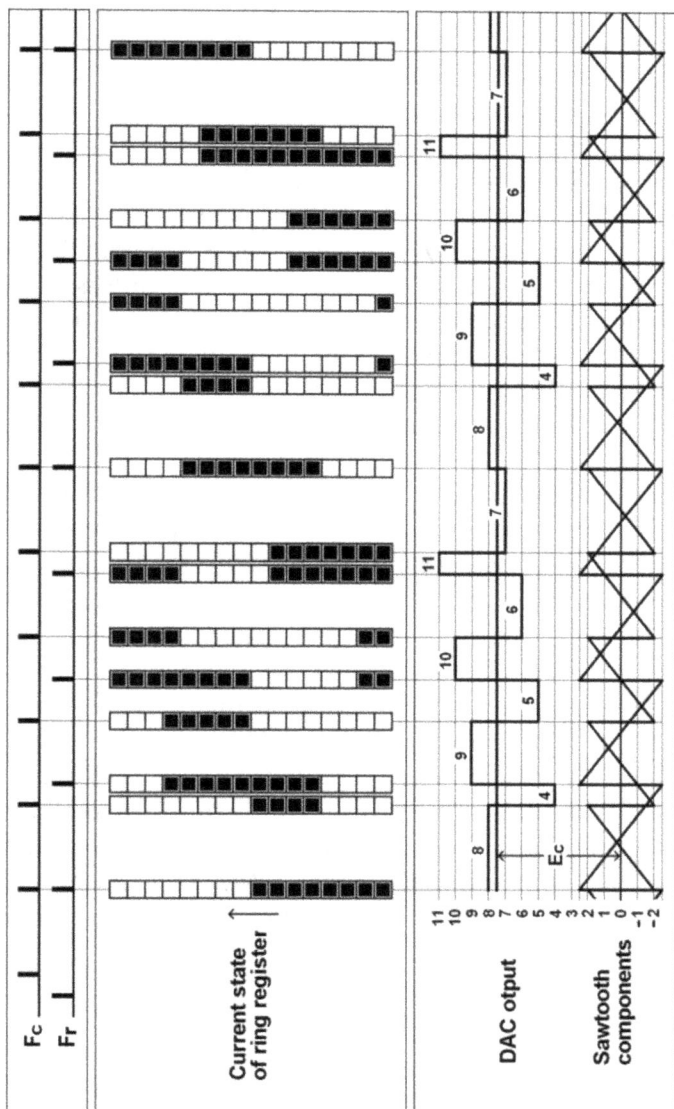

Figure 4.8 Diagrams explaining the work of the MFPD in Figure 4.7.

in the PLL, it does not create a multiplication of noise brought to its input.

In addition, due to the presence in the DAC many digits of equal weights, the requirements for the accuracy of each individual bit are reduced, or the accuracy of the DAC is improved with the initial inaccuracy of the digits, which contributes to a corresponding improvement in the signal spectrum.

However, although the described structure, when shown in generalized form, looks simple, it is not yet clear enough in what the construction of the impulse distributor will result. This issue can be circumvented by using two ring registers, reference and signal ones, clocked, respectively, with reference and signal pulses. Then the pulse distributor is completely excluded. In this case, the codes in one register and the other registers rotate in opposite directions. Also introduced are RS flip-flops by the number of register bits, operating from pulses in the register bits. Trigger outputs go to a segmented DAC.

Another question is about frequency resolution. If the task is to provide a fairly small frequency step, then the volume of registers may turn out to be above reasonable limits. Therefore, it is required to find a solution to how to construct the least significant bits, how to introduce a transfer from them into the register, and how to suppress the fractional noise that occurs in this case.

Although it is clear that the described idea can be successfully used to obtain a limited number of frequencies with high spectral purity, it can also be used in a more complex structure to form a coarse frequency step size that is filled, by summation in a separate PLL, with a fine steps obtained from a separate source, as described above.

Next, we will consider the variants of MFPD schemes, free from the noted drawbacks, and with improved characteristics in comparison with the previous versions.

References

[1] Bosselaers R. J., "Phase Locked Loop Including an Arithmetic Unit," US Patent 3,913,028, October 14, 1975, filed April 22, 1974.
[2] Koslov, V. I., et al., "Digital Phase Detector," Patent of Russia No. 875303, priority 12.02.80, in Russian.

[3]   Koslov, V. I., "Method of Phase Detection," *Radiotekhnika*, No. 4, 1980, in Russian.

[4]   Koslov, V. I., "Device for Digital Phase Detection of Pulse Sequences at Unequal Frequencies," Patent of Russia No. 1109872, priority 05.12.81, in Russian.

[5]   Koslov, V. I., "Frequency Synthesizers Based on Accumulators," *Electros-vyaz*, No. 2, 1988, in Russian.

[6]   Koslov, V., and N. Payne, "A New Approach to Frequency Synthesis," *Microwave Product Digest*, September 2011.

[7]   Makarenko, V., "Phase-Digital and Frequency-Digital Frequency Synthesizers, Part 1," *Electronic Components and Systems (EKiS)*, No. 11, November 2012, in Russian.

5

SYNTHESIZERS OF PHASE DIGITAL SYNTHESIZERS, AND PHASE DIGITAL SYNTHESIZERS WITH DELTA SIGMA MODULATION TYPES

5.1   THE IDEA OF PHASE SPLITTING

Suppose that in some system in which, along with the useful constant component, there is interference in the form of a pulse with a period of $T = 32$ conventional time intervals, as shown in Figure 5.1 (diagram A).

Let's shift the presented diagram by 1, 2, and 3 intervals (clocks) and sum up the original and shifted diagrams, each with a weight of $1/K$, where $K = 4$ (according to the number of diagrams); as a result, we get a B diagram. At the same time, it is clear that nothing could happen to the constant component, it will remain at the same level, and, therefore, it is not shown in the diagrams. As can be seen from this diagram, the amplitude of the received pulse decreased by 4 times, but the pulse width increased by the same amount, that is, its power remained the same. The C diagram shows the case in which the offsets are 2, 4, and 6 clocks. In this case, the interference power remained practically unchanged. If we choose shifts of 8, 16, and 24 clocks (D diagram), then the reducing the interference level turns out to be significant: moreover, its amplitude decreased by K times, and

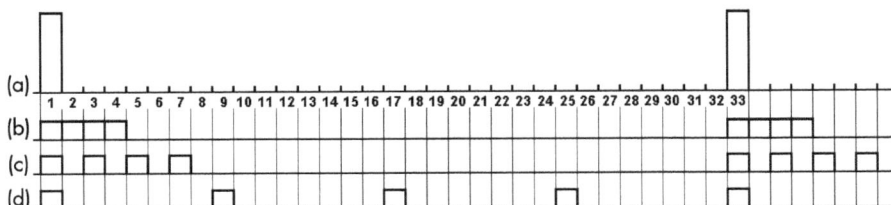

Figure 5.1 Clarifying the idea of splitting phases.

its frequency increased by K times (it is easier to filter it out). That is, for the idea to work effectively, the discreteness of the shifts must be equal to T/K.

The use of the described method for the purpose of frequency synthesis presupposes the presence of several partial phase detectors at the inputs of which there are pulse sequences that are compared in phase and are properly shifted in time relative to each other. Then, according to the described idea, a fraction $1/K$ of its total output voltage is taken from each partial detector.

It would seem that the idea is simple, universal, and suitable for dealing with interference in any frequency synthesis system. However, this is not quite true. In principle, it can be applied, for example, in a fractional-N PLL type synthesizer to reduce the fractional noise, and such attempts have been made [1, 2], but this raises practical problems. It is easy to imagine how long in time the process will be at the inputs of the partial detectors of the fractional-N PLL synthesizer, including in the variant with delta-sigma modulation, if needed to obtain a frequency step, say, 1 Hz. This is many millions of clocks, and therefore extremely long shift registers are required to obtain a significant positive effect. In addition, since the integer part of the division coefficient N changes, the length of the registers must also be changed, which is associated with a further significant complication of the structure. In practice, shifts of only a few clock cycles are used, which, naturally, is ineffective in reducing the level of spurs and fractional noise. Thus, in [3], it was shown that, when using 4 clock shifts, a gain in noise of the order of 3 dB is obtained, but with offsets from the signal significantly exceeding 10 kHz. At lower offsets, 10 kHz or less, there is practically no profit.

It should also be noted that in its simplest form, in synthesizers with integer division ratios, the idea of phase splitting is used to reduce the noise level by increasing the comparison frequency and by incoherent summing of noise [4]. It is also interesting to note that in this form the idea was patented as new [5] without reference to the primary sources [6, 7], where it was patented much earlier in a generalized form, and later developed in the works [8–13].

Next, we will consider the structures of the MFPD, in which the considered method operates in full force of its principal capabilities.

5.2 PREREQUISITES FOR USING THE IDEA

The idea of phase splitting can be used in two new types of synthesizers. For brevity, let's call them phase digital synthesizer (PDS) and phase digital synthesizer with delta sigma modulation (PDS-DSM). For them, the corresponding 2 versions of the MFPD include some of the blocks common to both options, as well as a number of other blocks with features for each of them. The structure common to both variants, which actually reflects the idea of phase splitting, is shown in Figure 5.2.

It includes identical reference and signal pulse distributors (in fact, these are phase splitters) operating on the block of partial phase detectors and then on the summer of the outputs of the detectors. Each of the phase splitters has, respectively, an input for a control code R or C and an input for a clock frequency F_r or F_c. They act in such a way that with each F_r or F_c clock, pulses, hereinafter referred to as split phases, in the amount of R or C arrive sequentially in time at the outputs of the phase splitters, as if they were closed in a ring.

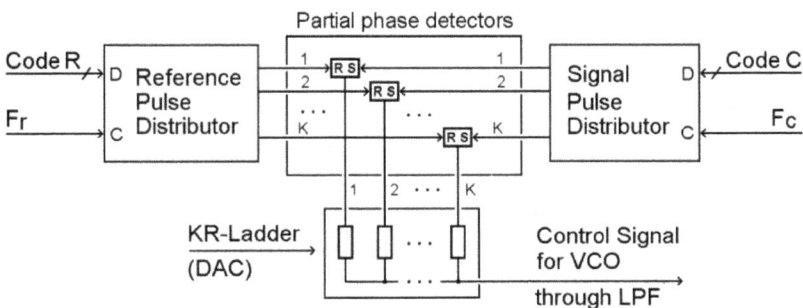

Figure 5.2 General structure for MFPD options.

The summer can be a KR ladder (with equal-weight resistors), or a segmented DAC can be used. At the output of the summer, a signal is generated that controls the frequency of the VCO through a lowpass filter.

The number of split phases K_r and K_c, respectively, in the reference and signal paths, can differ by an integer k times. Then the inputs of the partial phase detectors of the path with a greater number of phases are combined into groups of k inputs in each group and connected to the outputs of the phase splitter with a fewer number of phases. For example, if the number of phases in the reference path is $K_r = 32$, and in the signal path $K_c = 4$, then $k = K_r/K_c = 8$. In this case, a simple 4-bit ring counter can be used as a phase splitter in the signal path.

Figure 5.3 shows diagrams that explain the operation of the MFPD in comparison with the operation of a conventional phase detector: diagram A shows conventional PD, and diagram B shows MFPD. Phase splitters in MFPD can be of a rather large capacity Q and clocked with unequal frequencies F_r and F_c. However, for greater clarity of the comparison, the capacities in both paths are limited to a small value of $Q = 4$, clock frequencies are chosen equally, and the number of split phases equal $K = 4$.

An important fact follows from the consideration of Figure 5.3 that, in the MFPD, there is no need for frequency division to bring the frequencies to equality with subsequent phase comparison, which is inevitable in the case of a conventional system. The phase comparison takes place directly at the initial frequencies: the time position of each pulse of the signal sequence is controlled by reference pulses, and each reference pulse, through the MFPD, is used for this control, so there is no multiplication of input noise. It is also important that the partial detectors operate at lower frequencies, which provides a more accurate phase comparison.

However, the thesis about the absence of noise multiplication in the PLL needs additional explanation. This is practically true when the clock frequencies are approximately equal. If the signal frequency significantly exceeds the reference frequency, then, naturally, there is a multiplication in F_c/F_r times. From this follows a practical recommendation to maintain an approximate equality of clock frequencies and to increase the synthesized signal frequency by increasing the reference frequency. In the case when the possibilities of increasing

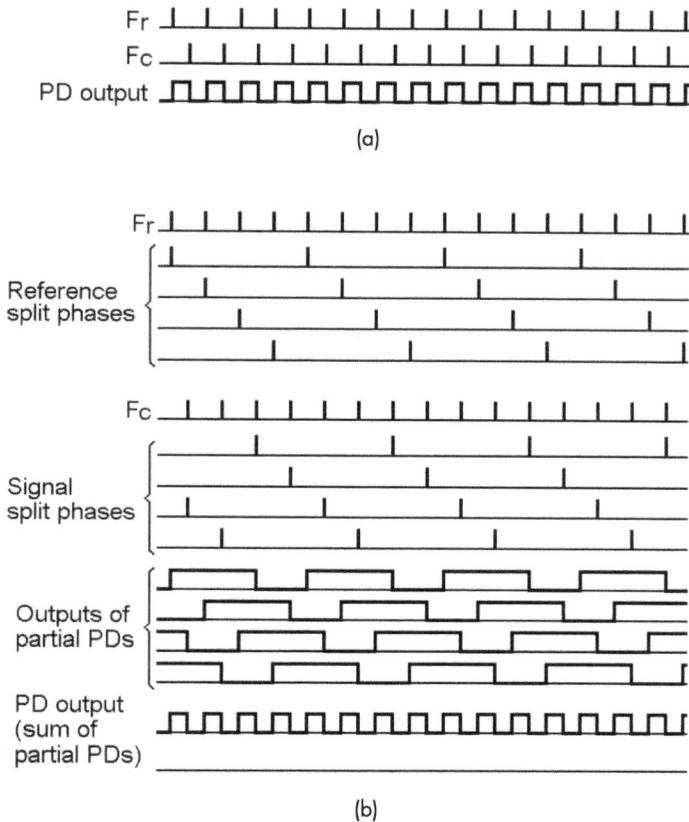

Fr

Fc

PD output

(a)

Fr

Reference
split phases

Fc

Signal
split phases

Outputs of
partial PDs

PD output
(sum of
partial PDs)

(b)

Figure 5.3 Comparison of (a) MFPD with (b) a conventional PD.

the clock frequencies are exhausted, for example, due to technological limitations in the phase splitters, then to increase the signal frequency at the PLL output, a prescaler should be included in the loop with the inevitable multiplication in the loop equal to the division factor of the prescaler.

5.3 MFPD IN VERSIONS FOR PDS AND PDS-DSM SYNTHESIZERS

5.3.1 MFPD for PDS

The MFPD circuit for a PDS is shown in Figure 5.4. The MFPD scheme looks quite symmetrical; it includes reference and signal paths that

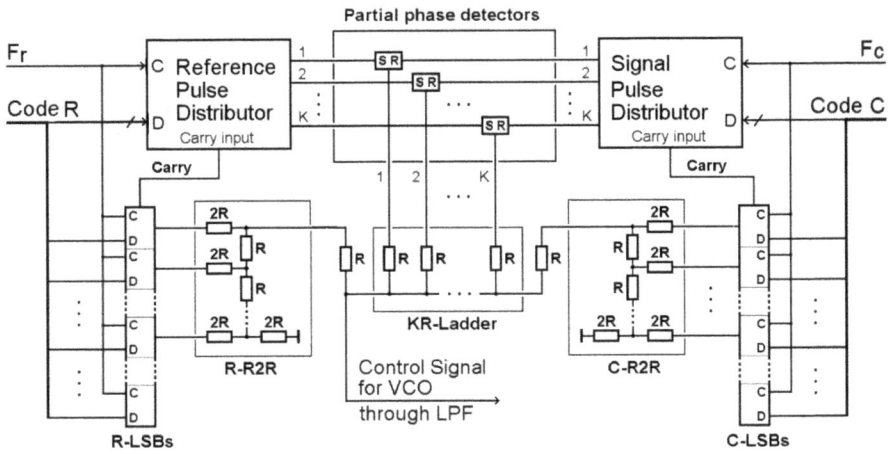

Figure 5.4 MFPD circuit for PDS.

are identical in appearance. However, as for the number of elements in the paths, there is no complete symmetry, which will be discussed below.

In the circuit of Figure 5.4, each of the paths, reference and signal, contains phase splitters controlled by the corresponding codes R and C. The phase splitters operate on partial phase detectors, which can be RS flip-flops, XOR elements, frequency-phase detectors with charge pump (FPD), and phase detectors of other types.

However, to provide the necessary frequency resolution, some split phases with a block of partial detectors may not be enough, since there are technological limitations on the number of split phases and the number of partial detectors equal to it. There is no problem in obtaining the number of phases, say, 32 or even 256, but to provide a frequency resolution of, say, 1 Hz, which is not such a strict requirement, it would take many thousands of them. This is beyond reasonable limits.

The problem is solved by the fact that, as shown in Figure 5.4, in the reference and signal paths, the accumulators of the LSBs, R-LSBs and C-LSBs, are connected to the buses of the corresponding control codes R and C and clocked with the same frequencies Fr and Fc. These accumulators form overflow (carry) pulses for the higher bits in pulse distributors (phase splitters) of the entire MFPD. In this way, due to the increase in the capacities of the paths, the desired frequency resolution is achieved.

In the general case, the numerical values of the control codes of the R-LSBs and C-MSBs accumulators are not multiples of the total capacities of the corresponding paths, which causes fractional interference at the synthesizer output. To suppress this interference, the R2R sections of the DAC, R-R2R and C-R2R are used, as shown in Figure 5.4. Their inputs are connected to the outputs of the higher digits of the corresponding accumulators R-LSBs and C-LSBs, and the outputs, through matching resistors, to the KR section of the DAC. The number of bits used to suppress the interference is determined by their achievable accuracy and is about 12 to 14 bits.

5.3.2 MFPD for PDS-DSM

First, let's turn to the history of the DSM method proposed by Wells [14, 15] from Marconi Instruments (now IFR) for using it for frequency synthesis. The method does not require analog components to suppress the fractional noise, which were necessary in the above fractional frequency division schemes.

The method itself was known much earlier, in the 1960s, and was used to build 1-bit DACs and ADCs [16–19]. In particular, Matsushita developed and used such a DAC for reading CDs in players. In addition, Nigel King of Racal Electronics [20] added a second accumulator to the RA1792 synthesizer, thus implementing second-order MASH much earlier than Wells. Brian Miller from Hewlett Packard also offered his own version of MASH [21]. Although this was later than the Wells patent, the version did not give any improvements, and the difference was, figuratively speaking, between $A + B$ and $B + A$.

The factors listed above have led to lengthy lawsuits between Marconi Instruments, Racal Electronics, Hewlett Packard, and others to establish the merits of each of the companies in the development of the method. However, these processes did not give results, and the companies decided to stop them, simply moving with their products to competition in the market.

However, the merit of Wells is undoubtedly the most significant, since it was he who first proposed the use of the method with the inclusion of all its possibilities, that is, with MASH of any order. In one of the articles from the company Marconi Instruments, for which he made this invention, he was called "the magician of the specialized laboratory."

Naturally, the first company to make extensive use of Wells's invention was Marconi Instruments, particularly in its 2030 and 2031 generators. In April 1995, it was awarded the Queens Award Technological Achievement in Fractional-N Systems. And after joining IFR, they continued to work in this direction, while simultaneously selling licenses for this technology to many other companies.

However, the path to a truly broad implementation of the idea turned out to be quite long. At least 20 years have passed since the publication of the Wells patent, when practically every company with access to integrated technology produces fractional-N PLL synthesizer chips of their own design. These are, for example, Analog Devices Inc. (ADI), Hittite (now in ADI), Texas Instruments, Synergy Microwave Corp., Maxim Integrated, Linear Technology Corp., and others. In these developments, Wells's idea did not undergo any significant changes and improvements, the mass of chips on the market is like twins, and it is difficult to imagine how each of the companies can withstand tough market competition.

The MFPD scheme for the PDS-DSM variant is shown in Figure 5.5. It differs in the presence of R-DSM and C-DSM blocks, which serve to suppress the fractional noise by DSM of the contents of the phase splitters. Each path contains an R-DSM or C-DSM accumulators for forming Pascal triangle sequences [12, 13], which are commands to add the corresponding numbers to the current states of the phase splitters, R-MSBs and C-MSBs, through the corresponding adders R-adder and C-adder. R-DSM and C-DSM accumulators may have each fewer bits required than R-LSBs and C-LSBs. As in the case of the PDS, it is limited by the achievable accuracy of the DAC (KR ladder) and can also be of the order of 12 to 14 bits.

Each of the described MFPD variants provides the synthesizer frequency F_c equal to

$$F_c = F_r R Q_c / (C Q_r)$$

where Q_r and Q_c are the total capacities of the reference and signal paths, including the capacities of the MSBs and LSBs, and R and C

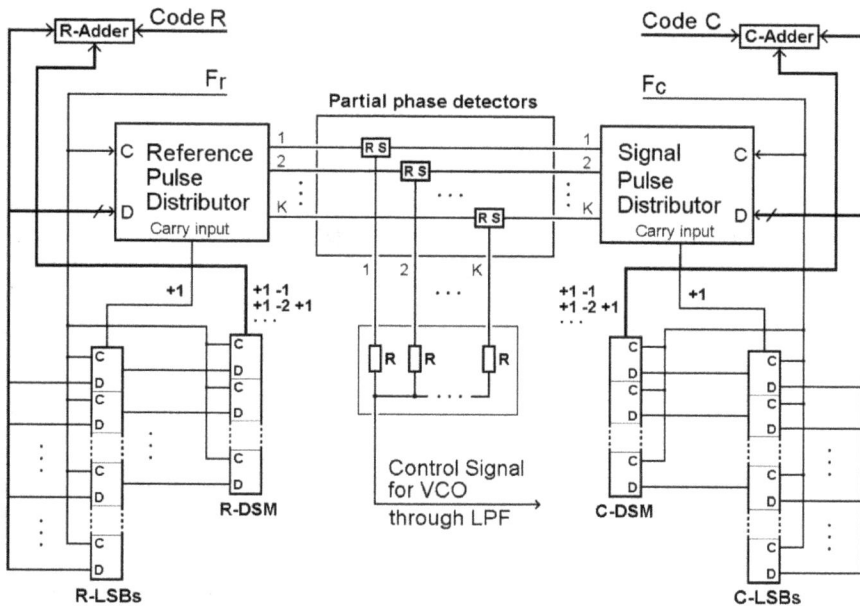

Figure 5.5 MFPD for PDS-DSM.

are the numerical values of the corresponding codes, represented by integers relative to the indicated capacities.

The step size dF of the frequency is

$$dF = F_r Q_c / (C Q_r)$$

As follows from the formula for F_c, it is possible to control the synthesizer frequency both by changing the value of the code R and code C. However, in order to provide the required, sufficiently small step size of the frequency, it is possible to limit ourselves to a correspondingly large capacity in only one, for example, in the reference path, and in the signal path it can be much smaller, up to the exclusion of the LSB block. Then it is still possible to obtain practically the same signal frequency with different combinations of the values of the R and C codes. This allows one to choose the most preferable combinations of these codes from the point of view of a minimum of fractional noise at the synthesizer output, in particular, to get rid of integer boundary spurs (IBS).

5.4   PHASE SPLITTERS

There are several options for phase splitter circuits for MFPD. Three of them are discussed below. These are options based on accumulators, digital adders, and logic elements.

5.4.1   MFPD with a Phase Splitter on Accumulators

Figure 5.6 shows the MFPD circuit, in which the functions of the phase splitter are performed by a set of $K = 4$ (as an example) accumulators. At the same time, the accumulators in their sum make up a block of more significant bits of a full accumulator, which, in addition, includes a block of less significant bits, which is also an accumulator.

The LSB block is connected to the MSBs by a carry chain, so that its overflow pulse in the form of a logical unit is fed to the carry receive input of each accumulator. Both blocks are clocked with the reference frequency F_r.

Accumulators in MSBs differ only in their initial operating conditions. If the accumulation process in the first one starts from 0, then the second one starts from the 1 state, the third one starts from the 2 state, and the fourth one starts from the 3 state. Accumulator overflow pulses are fed to the inputs of the partial detectors; each accumulator has its own detector. The other inputs of the detectors receive pulses from the signal phase splitter, which is used in this circuit as a ring counter clocked by signal pulses with a frequency F_c. This is chosen to simplify explanations about the operation of the MFPD. It corresponds to the case when the capacitance of the phase splitter in the signal path is $Q = 4$, and the numerical value of the code at its input is $C = 1$. Thus, it turns into the equivalent of a ring counter.

For the same purpose of simplification, the capacity of the accumulators was chosen to be small, only 2 binary digits in each accumulator. The value of the code at the full accumulator input is $R = 5$: one by one in MSB and LSB blocks. If considering the phase splitter with respect to the capacitance $Q = 4$, then the value of the R code is

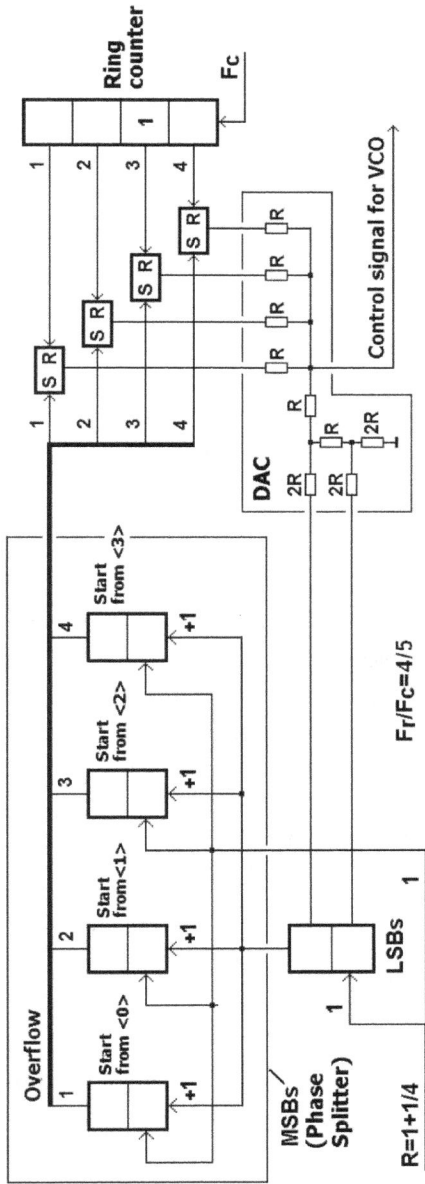

Figure 5.6 MFPD circuit with a phase splitter on accumulators.

$R = 1 + 1/4$. In this case, it is assumed, as an example, that RS flip-flops are used as partial detectors.

The outputs of the detectors are K high-order bits of the DAC. The LSBs of the DAC, built according to the R2R system, operate on pulses from the output of the LSB block, which, as will be shown below, serves to suppress the fractional noise. The DAC forms a voltage to control the VCO frequency.

Figure 5.7 shows the current states of the accumulator and the pulses at the output of the phase splitter. The values of the current contents of each accumulator at the moments of overflows are underlined in the figure. When the LSB accumulator overflows, it goes into the 0 state, and in this case, due to the transfer from the LSB block to the MSB block, one is added to the current content of the latter, as a result of which 2 pulses appear simultaneously on the adjacent outputs of the phase splitter instead of one. In this case, it turns out that the outputs of the phase splitter are, as it were, closed in a ring.

As can be seen, after each interval between pulses of 4 clocks in Figure 5.7, 4 intervals with a duration of 3 clocks follow. For greater clarity of this pattern, the interval of 4 bars is shaded. This indicates that the pulse processes at the outputs of the phase splitter are completely identical and are shifted in time relative to each other by $Q/K = 16/4 = 4$ clocks, where Q is the capacity of the full accumulator (including the block of less significant bits).

It is important to note that this pattern persists regardless of the width of the blocks' MSBs and LSBs. Thus, here is solved the problem that arose when trying to apply the idea of splitting phases in the fractional-N PLL synthesizer. It is not necessary to use multimillion-bit shift registers (see the previous section).

Figure 5.8 shows timing diagrams explaining the operation of the example of the MFPD in Figure 5.6.

The diagrams in Figure 5.8 are made for the case of a steady-state frequency ratio in the PLL system $F_r/F_c = 4/5$. The DAC output contains only a DC component, E_c, used to control the frequency of the VCO, and two sawtooth components, F_r and F_c, which are removed by the lowpass filter.

An important feature of the considered structure, as well as the previous versions of the MFPD, is the absence of frequency dividers to bring the frequencies of the reference and the signal to equality. Phase comparison takes place directly at the original frequencies. Therefore,

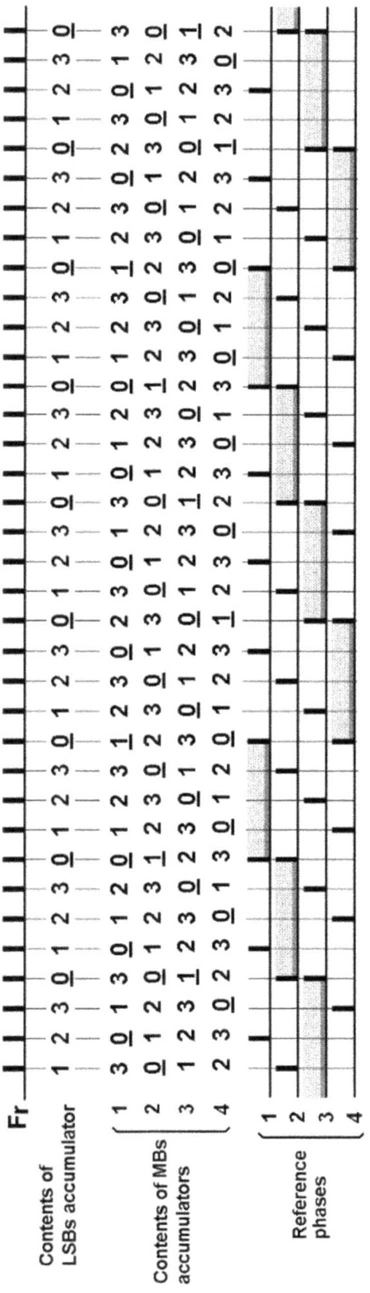

Figure 5.7 An explanation of the operation of the phase splitter.

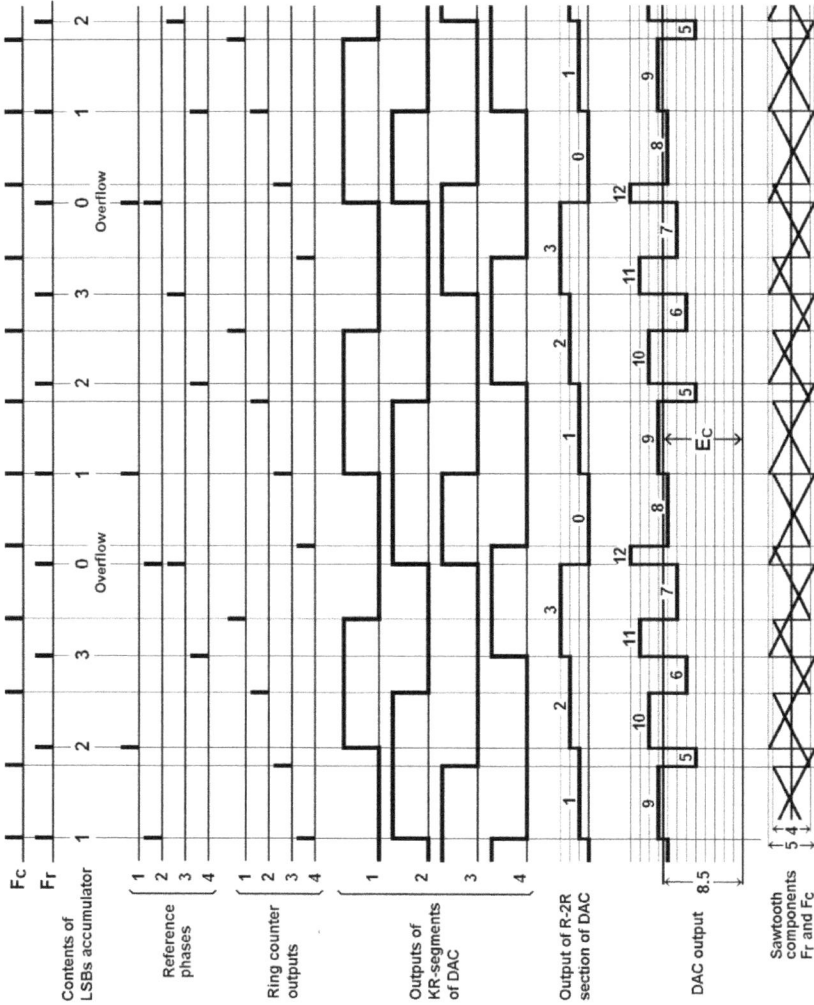

Figure 5.8 Timing diagrams explaining the work of the MFPD in Figure 5.6.

in the PLL system, there is no noise multiplication brought to the PD input, as is the case in fractional-N PLL synthesizers.

Another feature is the presence of many partial detectors and, accordingly, many bits in the segmented section of the DAC, and it can significantly increase its accuracy, which ultimately determines the spectral purity of the signal.

5.4.2 MFPD with Phase Splitter on Adders

Another option for constructing a phase splitter is also possible. It can be performed using digital adders as shown in Figure 5.9.

Here, as in the case of a phase splitter on accumulators, a simple example was chosen when the accumulator contains only two binary

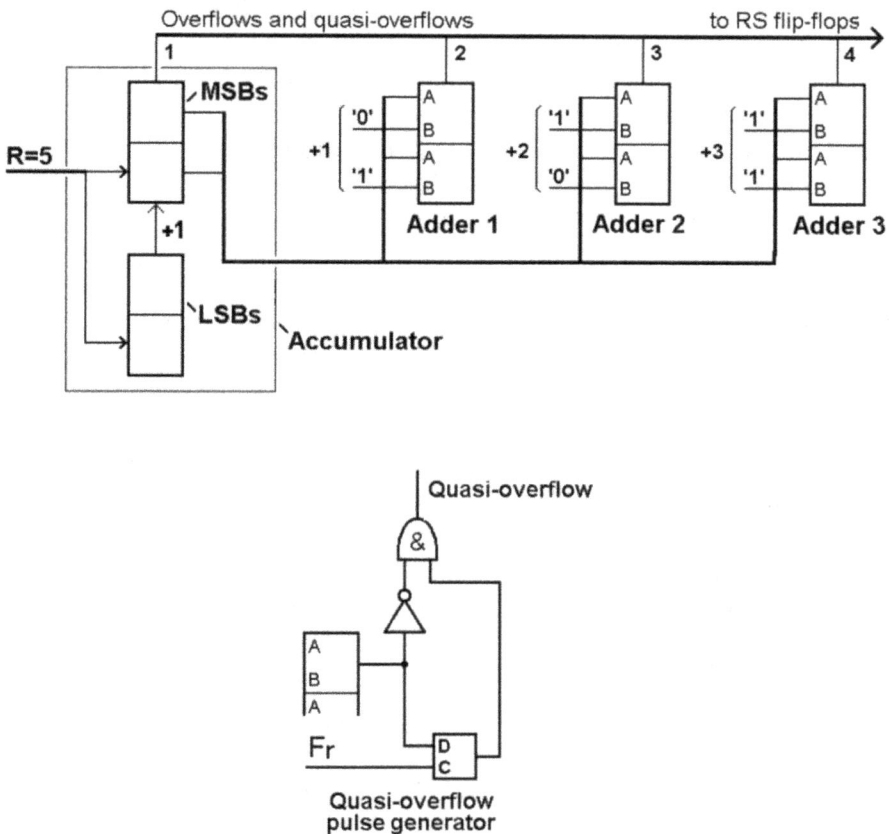

Figure 5.9 Phase splitter option on adders.

digits in the MSB and LSB blocks, that is, its total capacity is $Q = 16$. The numerical value of the code at the input to the accumulator is $R = 5$.

In the case when the number at the input B of the adder is less than or equal to the number at the input of the MSBs of the accumulator, then, at some points in time, the adder may not overflow. Therefore, a circuit is included in the MSB of the adder to fix the quasi-overflow moment. This is the moment when the MSB goes from the state 1 to the state 0. Such a circuit is shown in Figure 5.9. The output of the sum of the actual bit is connected to the input of the inverter and to the D input of the D flip-flop clocked by pulses F_r. The outputs of these elements are connected to the inputs of the AND circuit. At the moment when the state of the MSB changes from 1 to 0, two logical units appear at the input of the AND circuit, and, therefore, a pulse appears at its output. This is the quasi-overflow impulse. It is advisable to include the same circuit in the MSB of the accumulator in order to obtain a quasi-overflow pulse, so that its time position is consistent with the quasi-overflows of the adders.

Figure 5.10 shows the timing diagrams illustrating the operation of the scheme in Figure 5.9.

Figure 5.10 denotes and shows F_r (clock frequency), LSBs (the current contents of the block LSBs), MSBs (the current contents of the block MSBs and its overflow pulses), and MSBs+1, MSBs+2, and MSBs+3 (current states of adders and their overflow pulses, including those in the case of quasi-overflows). The values of the sums under quasi-overflows are underlined.

As can be seen from Figure 5.10, the arrangement of the pulses of the split phases is exactly the same as in the case of a phase splitter on accumulators. The pulse sequences in phases are shifted relative to each other by $Q/K = 4$ clocks, where K is the number of split phases. The considered option looks simpler than the previous one.

5.4.3 Phase Splitter on Logic Elements

A variant of a phase splitter using logic elements [6, 7] is shown in Figure 5.11.

To make it easier to understand its structure and operation, an example has been chosen when it is connected to the MSB block with

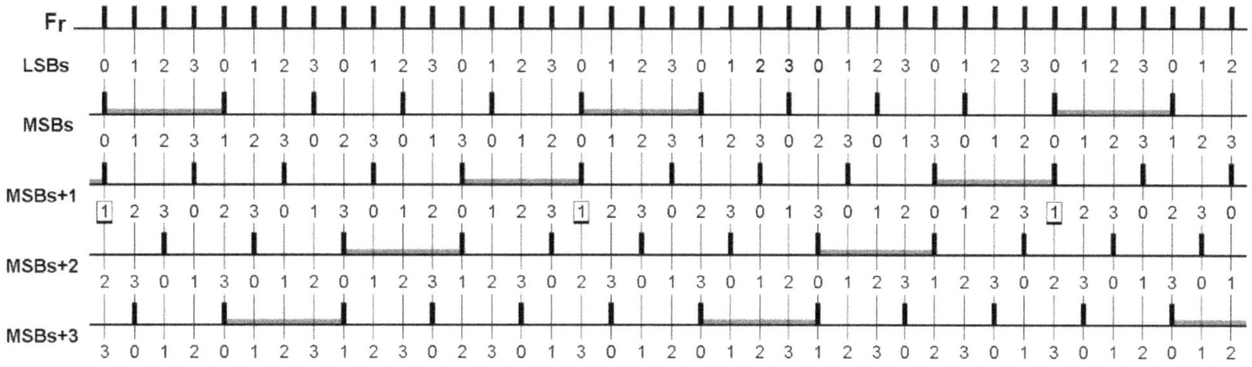

Figure 5.10 Diagrams explaining the operation of the phase splitter in Figure 5.9.

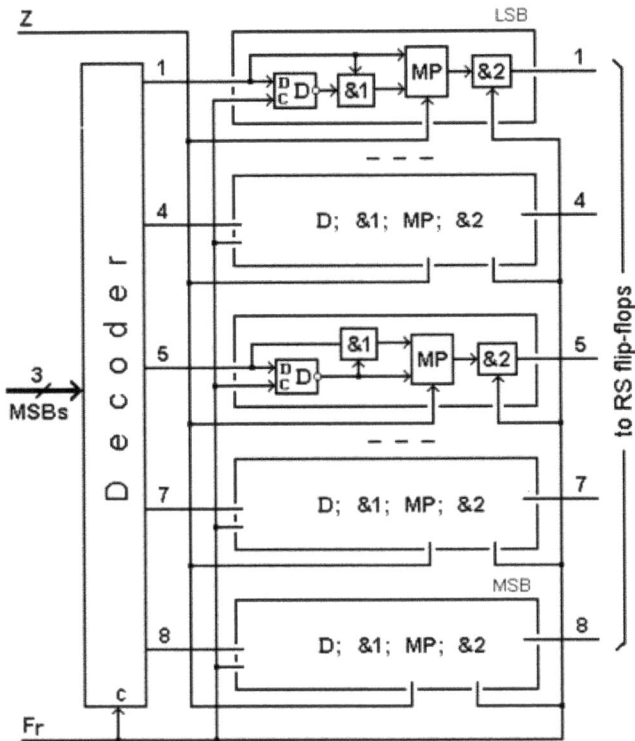

Figure 5.11 Phase splitter on logic elements.

a small number of bits, $k = 3$. The phase splitter includes a decoder and, accordingly, $K = 8$ logical circuits.

The decoder converts a binary code into a linear one, the number of units equal to the numerical value of the code. The decoder receives data from the MSBs of the main accumulator (see, for example, Figure 5.9) and generates input pulse sequences for the operation of logic circuits.

Below, the operation of the decoder is explained using the example shown in Figure 5.11, when it has a 3-bit bus at the input for the binary code and, accordingly, an 8-bit bus at the output for the linear code. The truth table for the decoder is presented in Table 5.1.

Each of the logic circuits of the phase splitter in Figure 5.11 contains a D flip-flop, AND1 element, 1-bit multiplexer MP, and AND2 element connected in series. In each of the logical circuits, the clock

Table 5.1

Decoder Input Codes	Codes at Decoder Outputs 1...8							
	1	2	3	4	5	6	7	8
000	0	0	0	0	0	0	0	0
001	1	0	0	0	0	0	0	0
010	1	1	0	0	0	0	0	0
011	1	1	1	0	0	0	0	0
100	1	1	1	1	0	0	0	0
101	1	1	1	1	1	0	0	0
110	1	1	1	1	1	1	0	0
111	1	1	1	1	1	1	1	0

input C of the D flip-flop and one of the inputs of the AND2 element are connected to the reference frequency bus. The control inputs of the multiplexers are combined to form the control input Z of the phase splitter connected to the overflow circuit shown in Figure 5.9.

In half of the logical circuits, namely in circuits, the inputs and outputs of which are numbered 1..4 (let's call them circuits 1..4) and representing the less significant bits of the code at the D input of the phase splitter, the input of each of the logical circuits is connected to the D input of the D flip-flop, one of the inputs of the element AND1, and one of the inputs of the multiplexer MP. The inverse output of the D flip-flop is connected to another input of the element AND1, and the output of the multiplexer MP is connected to another input of the element AND2. The output of each circuit AND2 is one of the four output bits of the phase splitter.

The circuits of the other half of the logic circuits, the inputs and outputs of which are numbered as 5..8 (let's call them circuits 5..8), are built basically the same way, except that the AND1 element is located between the input of the logic circuit and the multiplexer, and the output of the D flip-flop is connected to one of the inputs of the multiplexer MP. The output of each element AND2 is one of the four remaining bit outputs of the phase splitter.

The phase splitter works as follows. D flip-flops of all logic circuits store the value of the linear code obtained in the previous cycle, and the AND1 elements subtract the previous code value from its current one. The resulting difference is transmitted through multiplexers to the AND2 elements, which allow (in the presence of ones in the received code) pulses F_r to pass to the output of the phase splitter and then to the inputs of the partial phase detectors.

The results of the subtractions are used in all clocks up to the moment the accumulator is full, and, consequently, until the decoder is full. Until this moment, there is no accumulator overflow pulse at the control inputs of the multiplexers MP, and each multiplexer allows pulses F_r to pass from the output of the corresponding element AND1 to the corresponding element AND2. As soon as the high bit of the accumulator passes from the state 1 to the state 0 (the moment of overflow of the accumulator), a pulse appears at the control inputs of the multiplexers, which disconnects the outputs of the AND1 elements from the inputs of the AND2 elements.

Simultaneously, in logical circuits 1..4, corresponding to less significant bits of the input code, multiplexers connect decoder outputs to AND2 elements, and in logical circuits 5..8 connect outputs of the corresponding D flip-flops. This is necessary to start the next decoder filling cycle in order to ensure the rotation of units at the multiplexer outputs, as it were, closed in a ring.

Under the action of a logic level at the multiplexer output applied to one of the inputs of the AND2 element, the latter enables or disables the passage of F_r pulses to the input of the corresponding partial phase detector.

For example, let's assume that the control code at the input of the MSB block of the accumulator is equal to $R_1 = 3$ (while the capacity of the latter is $k = 3$ digits, that is, $K = 8$), and the code at the input of the LSB block is $R_2 = 0$, that is, the block is disabled. Then, in accordance with the principle of operation of the phase splitter described above, the variable code $R_1(i)$ (i is the clock number) at the data input of the phase splitter is converted into the process of distributing F_r pulses over the outputs of the phase splitter as shown in Table 5.2, where the digit 1 means the presence of a pulse at the corresponding output bit of the phase splitter or, equivalently, at the input of the corresponding partial detector. Hereinafter, numbers corresponding to accumulator overflow states are enclosed in parentheses.

As can be seen from Table 5.2 (top), in each clock, pulses appear at the inputs of three adjacent detectors, and, with each clock, the pulses are shifted to the next three detectors, moving along the ring.

Table 5.2

R1=3; R2=0

Clock number i	1	2	3	4	5	6	7	8	9	10	11	12	13	14	...
R1(i)	4	7	[2]	6	[1]	4	7	[2]	5	[0]	3	7	[2]	5	...
Flip-flop number and the presence of a pulse at its input — 1		1			1		1				1		1		
2	1	1			1			1			1		1		
3	1			1	1			1				1	1		
4	1			1			1	1				1			
5		1		1			1			1		1			
6		1			1		1			1			1		
7		1		1				1	1			1			
8			1		1			1			1		1		

R1=3; R2=1

Clock number i	1	2	3	4	5	6	7	8	9	10	11	12	13	14	...
R2(i)	5	6	7	[0]	1	2	3	4	5	6	7	[0]	1	2	...
R1(i)	4	7	[2]	6	[1]	4	7	[2]	5	[0]	3	7	[2]	5	...
Flip-flop number and the presence of a pulse at its input — 1		1		1			1				1		1		
2	1	1			1		1				1		1		
3	1			1	1			1			1			1	
4	1			1	1			1				1		1	
5		1		1			1		1			1		1	
6		1		1			1			1		1			
7		1			1		1				1		1		
8			1		1			1			1		1		

Table 5.2 also shows the case when the R_2 code at the input of the LSB block is not 0 (bottom). In this example, $R_2 = 1$, and, as mentioned above, the number of bits of the LSB block is 3. Then there is a process $R_2(i)$ in this block, and in certain clocks the overflow pulse of this block is transmitted to the block MSBs. As a result, the number of pulses at the output of the phase splitter in these clocks increases by one (see clocks with numbers $i = 4$ and 12).

Here, for simplicity of presentation, we limited ourselves to small bit widths of MSBs and LSBs, assuming that it is easy to imagine processes in a real multibit MFPD. From a consideration of the operation of this version of the phase splitter, it is also clear that the result of its action is exactly the same as the other two variants given earlier.

5.5 STATIC CHARACTERISTICS OF MFPD

The static characteristics of the MFPD have the peculiarity that their slope is generally not the same for the reference and signal paths. Let's show this by calculations.

The phase extent of the static characteristics for the reference and signal paths is determined by the comparison frequencies in the partial phase detectors. These frequencies are equal, on average, for the reference path $-RF_r/Q_r$, and for the signal path $-CF_c/Q_c$, where Q_r and Q_c are the capacitances of the pulse distributors, respectively, of the reference and signal paths, and R and C are the corresponding numerical values of the control codes.

In the steady state, that is, in the state of synchronism in the PLL, these frequencies are equal, but the frequencies F_r and F_c are generally different, and therefore the ratios R/Q_r and C/Q_c differ by the same amount. This means that the phase length of the characteristic of the reference path is $2\pi Q_r/R$, and for the signal path $-2\pi Qc/C$. This gives the expressions for the steepness of the static characteristics S_r and S_c for the reference and signal paths, respectively:

$$S_r = U_{max}R/(S\pi Q_r) \text{ and } S_c = U_{max}C/(2\pi Q_c)$$

where U_{max} is the full scale of the voltage at the output of the MFPD.

Figure 5.12 clearly shows the confirmation of the formulas obtained. The figure illustrates the case when $Q_r = Q_c = 8$; $R = 3$, $C = 2$ (i.e., the frequency ratio is equal to $F_c/F_r = 3/2$). The diagram A refers to some initial position of the pulses F_r and F_c; diagram B refers to the case when the pulses in the reference path are shifted to the right relative to the initial ones by half a period F_r, that is, by π radians; and diagram C refers to when the pulses in the signal path are shifted to the left, relative to the initial ones, by half a period F_c, that is, also by π radians. The full scale of the voltage at the DAC output, in accordance with the capacitances Q_r and Q_c, is $U_{max} = 8$ conventional units.

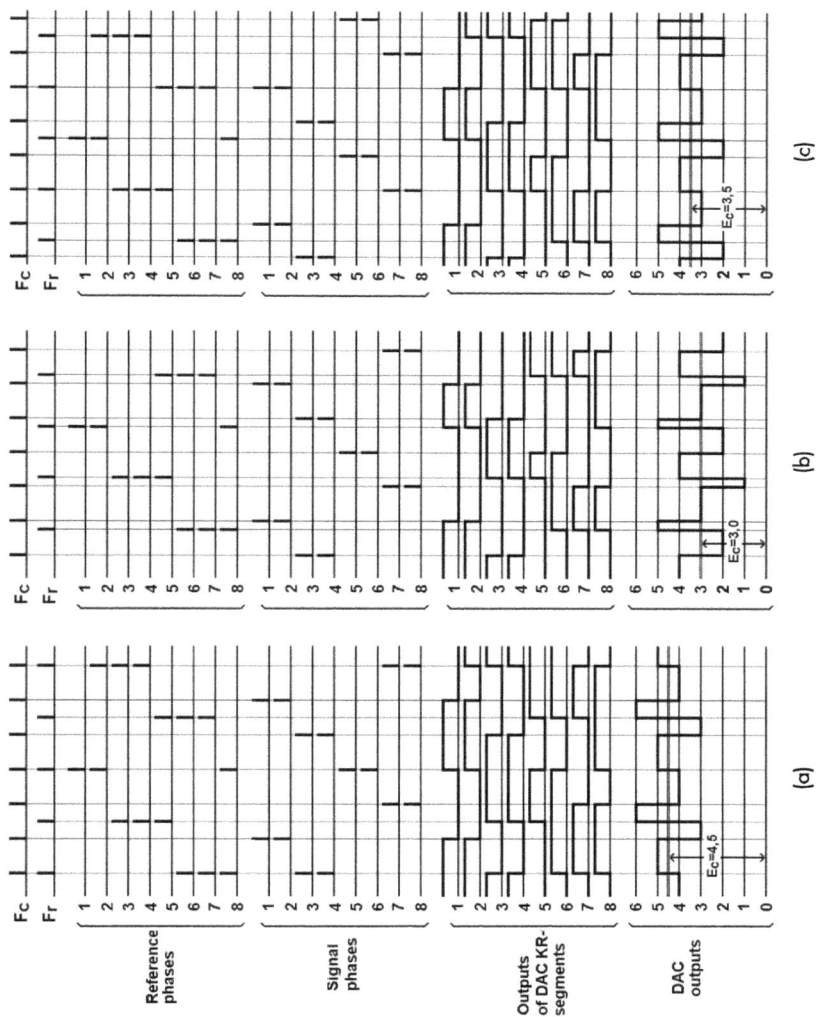

Figure 5.12 The determination of the static characteristics of the MFPD. (a) Initial positon of pulses F_r and F_c, (b) pulses F_r are shifted half a period from right, and (c) pulses F_c are shifted half a period to left.

In diagram B, the constant component E_c has changed by $dEc = 1.5$ units. If we now use the above expression for the slope S_r, we get the same value 1.5.

In diagram C the constant component E_c has decreased by $dEc = 1$. Using the expression for the slope S_c, we get the same number, which is reflected in the figure. In the first case, for the reference path, the slope of the static characteristic is $S_r = 3U_{max}/16\pi$, and in the second case, for the signal path, $S_c = U_{max}/8\pi$.

The above examples confirm the validity of the obtained formulas for the steepness of the static characteristics of the MFPD in the reference and signal paths.

It is clear that this or that expression for the slope of the characteristic is used depending on the structure of the PLL model: to which point of the model an external influence is applied, and where the system's response to this influence is expected. At the same time, one must also take into account that the characteristics are rigidly connected with each other through the frequency ratio $S_c/S_r = F_r/F_c$; knowing one of them, it is easy to calculate the other.

It is important to note the following here. As can be seen from Figure 5.12, each pulse at the inputs of the partial detectors corresponds to the initial pulse at the inputs of the MFDD, that is, the input pulses operate continuously, without gaps, and this means that there is no frequency division in the PLL system.

5.6   THE OPERATION AREA OF THE STATIC CHARACTERISTICS OF THE MFPD

Let's refer to Figure 5.13. On diagram A, the phase difference is such that the pulse duration in the partial detectors is the minimum possible, and the MPPD output is 1.5 units at a full DAC scale of 8 units. Diagram B refers to the case when the pulses in the partial detectors are of the maximum possible duration, and then the DAC output is 6.5 units.

Thus, $R/Q_r = 3/8$ area of the characteristic is lost for operation. For practice, an R/Q_r ratio around the mean of 1/4 can be recommended. Then the slope of the characteristic will be of the order of $S_r = U_{max}/(8\pi)$, and the length of the characteristic in amplitude will be of the order of $3/4 \ U_{max}$.

(a)

(b)

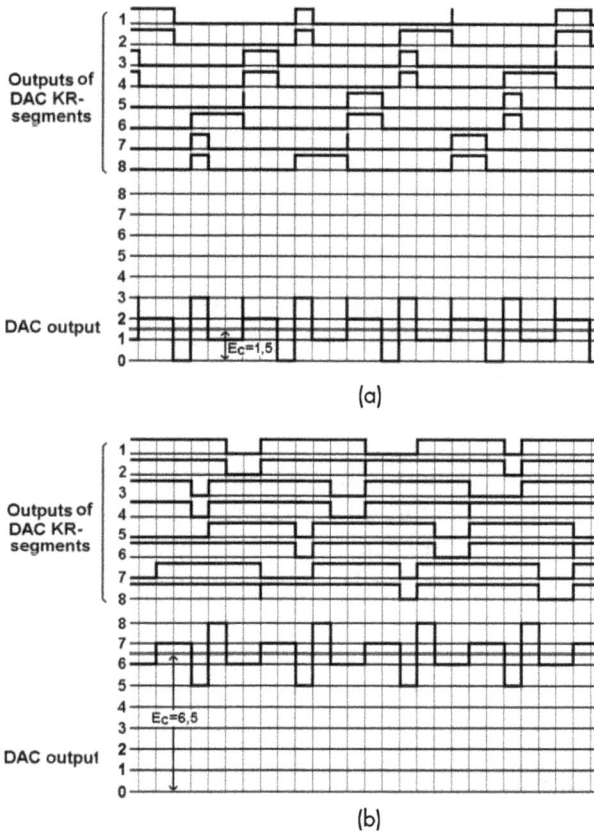

Figure 5.13 The determination of the operation area of the characteristics of the MFPD: (a) lower limit level E_c and (b) upper limit level E_c.

The results obtained above relate to the case when the ratio $Q_c/C = 4$ in the signal path is an integer, and therefore a ring counter can be used as a phase splitter, the output process of which is periodic and does not affect the length of the linear, working area of the static characteristic of the MFPD. If the phase splitter of the signal path has the same structure as in the reference path, and the Q_c/C ratio in the general case is not an integer, then the picture of the process at the output of the MFPD becomes much more complicated. However, it is quite clear that in this case the principle of superposition is valid, and in this case the length of the linear area of the characteristic will additionally decrease by the C/Q_c of U_{max}. If, for example, the ratios R/Q_r

and C/Q_c are of the order of 1/4, then the linear area will be reduced to half of the U_{max} scale. In this case, the switching frequency of the DAC segments is approximately 4 times lower than the reference frequency, which improves the DAC accuracy.

5.7   SIGNAL SPECTRA OF SYNTHESIZERS OF PDS AND PDS-DSM TYPES

As noted above, frequency synthesizers using the MFPD are PDS. So far, we have considered idealized versions of circuits in which the DAC did not have errors, and therefore, the fractional noise in the PDS version was absent, and in the PDS-DSM version it was negligible.

In reality, in order to get an idea of the spectral purity of the synthesizer signal, it is necessary to know these errors. Naturally, the greatest contribution to the degradation of the signal spectrum of a synthesizer with an MFPD in the PLL is made by the inaccuracies of the high-order bits of the DAC, namely its KR segments. Therefore, the inaccuracies of R2R sections due to their significantly lower weight can be ignored. It is not easy to calculate the effect of inaccuracies of the digits, when there are many of them (for example, 32), on the spectrum, but some dependencies are known.

For example, when the inaccuracies of adjacent digits, say, the first and second, are equal and opposite in sign, then the level of the first harmonic of the interference decreases by 4.7 dB compared to the inaccuracy of one digit. In the case of anti-phase bits, for example, 1 and 17, (when $q = 32$) if their inaccuracies in magnitude and sign are equal, there is no interference from them at all.

In general, we have to resort to statistical calculation methods. From the experience of Analog Devices, it is known that the inaccuracy of one KR segment can be reduced to a value not exceeding 0.1% of its weight. Let's further assume that the inaccuracies of the bits are distributed according to the normal law with an rms value of 0.1%.

One example of such a distribution is shown in Figure 5.14, where there are the bit numbers and the percentage of their inaccuracy.

In his article [22] about fractional noise in PDS (with MFPD), on the Simulink-MATLAB model with a variety of variants of this type of distribution of inaccuracies, Romanov revealed a simple regularity: the level of noise in the spectrum of each of the options increases no

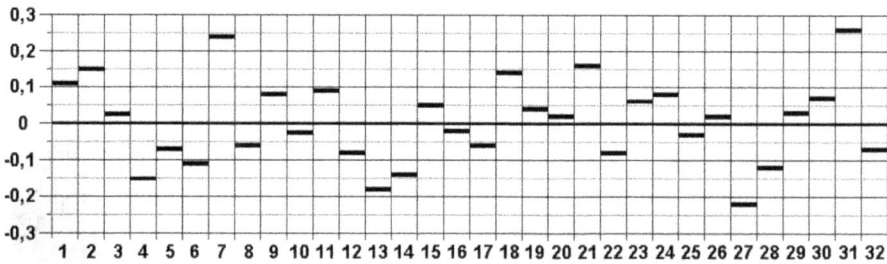

Figure 5.14 An example of distribution of inaccuracies of KR segments of DAC.

more than by 12 dB in comparison with the spectrum with the same error, but in one segment of DAC.

Therefore, calculations can be carried out for one segment of DAC, which is much simpler, and for this a special program has been developed that allows significantly, in comparison with Simulink-MATLAB, to increase the estimated number of accumulator segments and the range of offsets from the signal carrier frequency.

In connection with the above, we present the calculations of the spectra of phase noise due to fractional interference with an error of one of the KR segments of the DAC equal to 0.4%, assuming that this corresponds to an rms inaccuracy of 0.1% with a normal distribution of inaccuracies over all segments.

Typical noise spectra at the output of a PDS corresponding to this approach are shown in Figure 5.15. The following parameters are selected for the MFPD and PLL. The total number of MFPD bits is 19, of which 5 are senior ones, having an output through a phase splitter to the KR section of the DAC, which includes 32 segments. A 4-bit ring counter is used as a phase splitter in the signal path, and a prescaler with a division factor of 4 is included in the PLL system. The reference frequency is 1 GHz. There is no lowpass filter in the PLL. Its influence on the spectrum can be taken into account separately.

The spectrum A corresponds to the code $R =$ 01000.00000000000001, at which the signal frequency is $F_c =$ 4.000030518 GHz. This case can be called, by analogy with the fractional-N PLL synthesizer, as IBS, when the value of the R code is fractional and closest to its integer value. The integer value of the code is the one by which the total capacity of the accumulator is divided without a remainder. All other values are fractional.

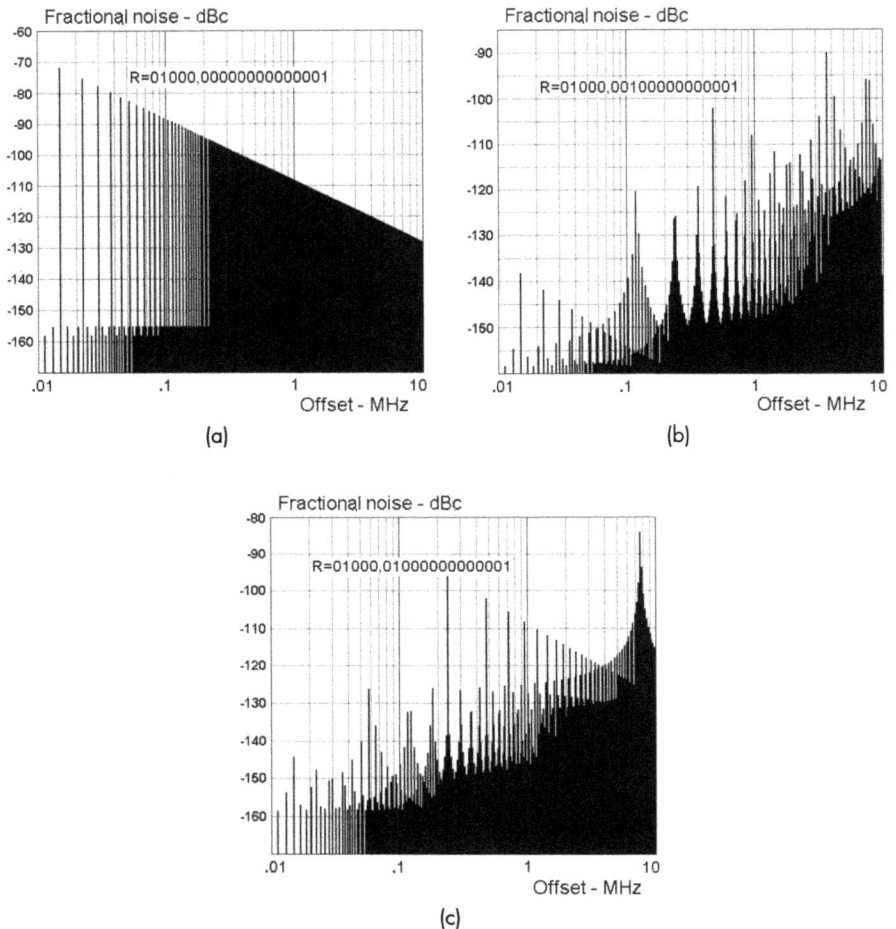

Figure 5.15 Fractional noise at the output of the PDS.

The structure of the A spectrum is determined by the position of one in the LSB of the control code R. In this case, the noise appears at the DAC output in the form of a sawtooth component modulating the VCO signal. The interference frequency is calculated based on the number of zeros in the R code between the major and minor ones. The more zeros, the lower the frequency. The physical meaning of this lies in the process of accumulation of the lowest unit in the accumulator, namely, in the duration of this process, when there will be a transfer from the lowest bits to the bit corresponding to the highest unit at the input of the accumulator. Therefore, the first harmonic of the interference in the spectrum is closest to the signal and has the highest level for all possible combinations of ones and zeros in the R code. Higher harmonics decrease monotonically in accordance with the sawtooth nature of the interference. It is clear that it is desirable to exclude the case of IBS from use in the practical application of the synthesizer. It will be explained below how this can be done.

Here we have simplified the structure of the PDS by assuming that a noninterfering ring counter is installed in the signal path. This is done with the superposition principle in mind. If there is also a phase splitter in the signal path, similar to the reference path, then its interference spectrum, calculated in exactly the same way, will be superimposed on the interference spectrum of the reference path. However, the bit width of the signal distributor of pulses can be small, so that the noise from it turns out to be high-frequency enough to be filtered out without significantly narrowing the PLL bandwidth, that is, practically without reducing the synthesizer speed. Therefore, the choice of the most successful combination of codes R and C, from the point of view of the spectral purity of the signal, is possible, including the exclusion of the IBS case.

The   B   diagram   was   obtained   with   the   code R = 01000.001000000000001 giving the signal frequency $F_c = 4.062530518$ GHz, and can be called the 1/64 IBS case, since the signal frequency is offset by 1/64 from the frequency determined by the integer value of the R code.

The   spectrum C corresponds   to   the   code R = 01000.01000000000001, and the signal frequency equal to $F_c = 4.125030518$ GHz. This case can be referred to as 1/32 IBS, according to the offset of the signal by 1/32 of the frequency mentioned above.

In the last two cases, there are clearly pronounced discrete components in the spectrum. Their position on the offset axis is determined by the position of the units in the R code. Lower-frequency interference components, determined by the position of the less significant units of the code, are the modulating factor for the higher-frequency components, determined by the position of the higher ones. Due to this, the higher-level subcarriers arises, around which the side bands of low-frequency interference of a lower level are grouped. As a result, the level of spectral components turns out to be significantly lower than in the case of IBS. In addition, they are quite far from the carrier and therefore can be filtered out by the PLL.

Figure 5.16 shows the signal spectrum of PDS-DSM synthesizer when 2 accumulators in series are included in the DSM block (MASH-3). Otherwise, the parameters of the MFPD and PLL, as well as the R code values for the corresponding A, B, and C diagrams, remain the same. In contrast to PDS, PDS-DSM spectra are smoother, and fractional noise increases monotonously, mainly with increasing signal offset. It is also easy to see that the case of IBS in PDS-DSM (diagram A) is significantly less pronounced than in the case of PDS.

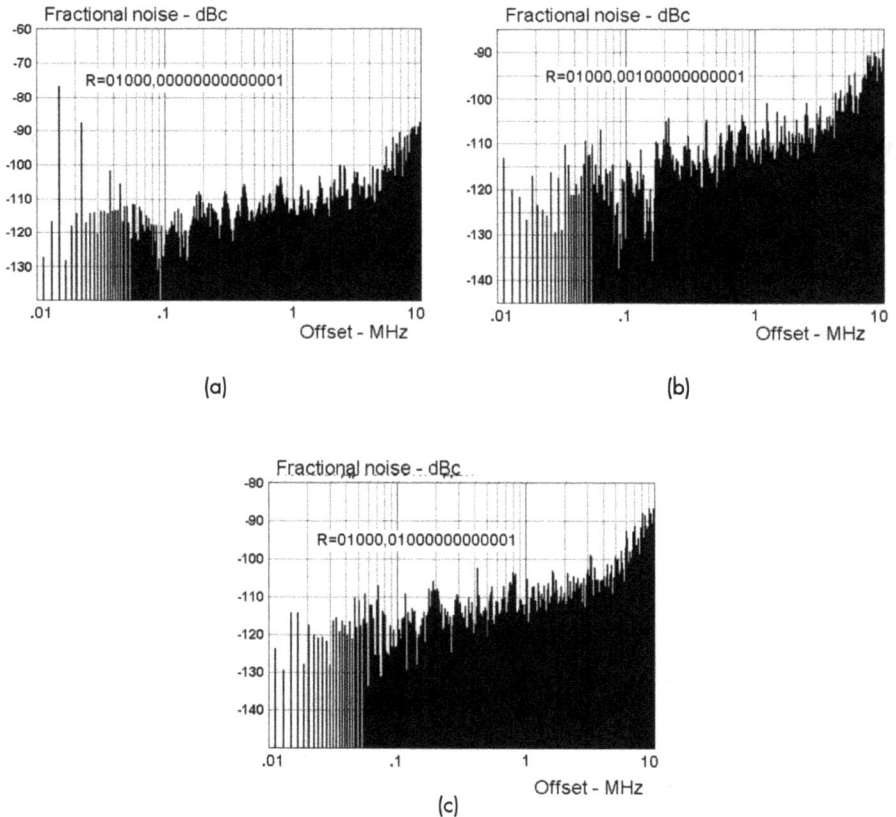

Figure 5.16 Fractional noise spectrum of the PDS-DSM.

The timing mismatch of the DAC bits also affects the quality of the signal spectrum. As an example, Figure 5.17 shows the plots of the total power of the fractional phase noise in the PLL bandwidth for the case of 1/32 IBS, when in one of the KR segments the pulses are shifted by 40 picoseconds. (For Analog Devices products, this figure does not exceed 10 pS.) The reference frequency is $F_r = 1,000$ MHz, and the signal frequency is $F_c = 4.125030518$ GHz $(R = 01000.01000000000001)$.

Calculations are performed for both PDS and PDS-DSM (MASH-3) versions. Figure 5.17 also shows the spectrum for PDS-DSM in the absence of time error $(d_t = 0)$. There is no similar spectrum for the PDS variant, since in this case there is no fractional interference in it at all.

As can be seen from the diagrams above, for some PLL bandwidth values, the PDS version has some advantage in spectral purity over the PDS-DSM version. However, for the worst cases, their capabilities are about the same.

Comparison of the spectral diagrams in Figure 5.17 leads to the conclusion that the capabilities of the PDS and PDS-DSM variants

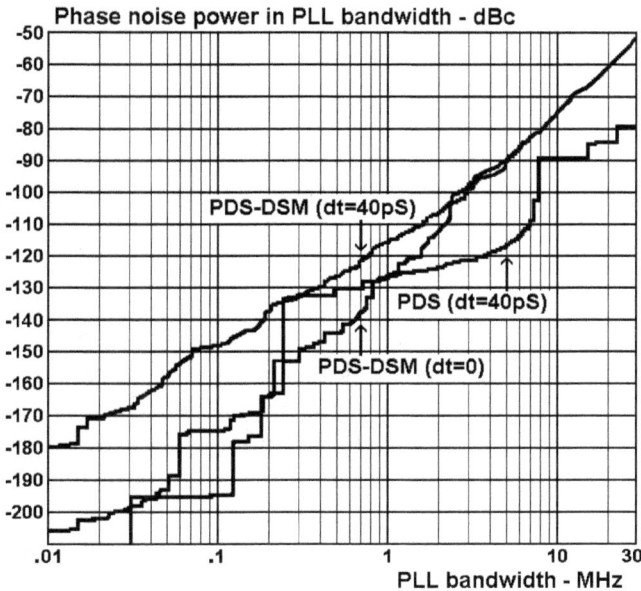

Figure 5.17 Fractional noise at time mismatches in the DAC.

in ensuring the spectral purity of the signal, in the presence of time mismatches, are quite high and of about the same order of magnitude.

5.8 REDUCTION OF POWER CONSUMPTION

The power consumption of the PDS-DSM can be significantly reduced. This is possible for the cases of using the MFPD, when it is not required to provide an extremely wide bandwidth of the PLL system and, therefore, the speed of frequency switching.

Let's assume that the required frequency resolution is provided by a sufficiently large capacity of the reference path, while the capacity of the signal path is limited to several bits, as mentioned above. Then the clock frequency of R-LSB and R-DSM blocks in the PDS-DSM version can be reduced several times. As a result, the power consumed by both these units and the MFPD, as a whole, decreases. The gain is quite significant, since a significant proportion of the circuit elements are in these blocks.

In this case, in order to maintain the same efficiency of fractional noise suppression, the duration of the carry pulse from the R-LSBs to the R-MSBs, as well as the duration of the overflow pulse of the R-DSM, must be shortened to the period of the reference frequency F_r.

Due to the fact that the clock frequency of the R-LSB block has decreased, let's say, by $M = 4$ times, the weight of the bits of this block has decreased by the same amount. This means that its capacity, as it were, increased by 2 digits and, accordingly, the frequency resolution improved. To leave the frequency resolution the same, the 2 LSBs of this block need to be removed.

Another peculiarity is that with lowering the clock frequency of block R-LSBs, for example, by 4 times, there appeared two additional more significant bits in this block to which there is no access for tuning code R because the bits are absent as the physical ones. Due to this fact, dead zones appear in the frequency range of the synthesizer, in which some number of frequencies cannot be provided. To cover these zones, 2 bits need to be added into the C-LSB block in order to get possibility for choosing the corresponding combinations of codes R and C.

Note that, if the value of ratio M is not as 2^m, then the number of the mentioned bits for excluding from the R-LSB block needs to be nearest to but less than 2^m in order not to worsen previous frequency resolution, and the number of bits for adding into the C-LSB block needs to be nearest to but more than 2^m in order not to lose the parts of the frequency range. For example, if $M = 5$, then 2 bits in the R-LSB block may be excluded and 3 bits have to be added into the C-LSB block.

In Figure 5.18, for comparison, the spectrum diagrams (spectral density of phase noise) of the version with R-LSB and R-DSM blocks are depicted with and without the divisor of $M = 8$. While the reference frequency is equal to $F_r = 1,000$ MHz, the output frequency of synthesizer equal $F_c = 1,015.6326$ MHz. The number of split phases is 64. In the DSM block, there are included 2 accumulators (MASH-3), and a tuning code R is $R = 010000.01000000000001$ (1/64 IBS). The inaccuracy of one KR segment of the DAC is 0.4%, which, as noted above, corresponds to 0.1% of the rms inaccuracy with the normal law of its distribution over bits.

Figure 5.18 shows how the diagram of noise is changed with including the divider. If considering that the maximum possible PLL bandwidth is determined by the point on the axis of offsets where noises begin to rise sharply because of the action of DSM, and they should be effectively suppressed, then, as it follows from the diagrams, PLL bandwidth with a frequency divider of $M = 8$ is decreased by 2 orders as compared with the variant without divider. The level of noise in the PLL bandwidth is increased by approximately 6 dB. Thus, it is the recompense for reducing power consumption.

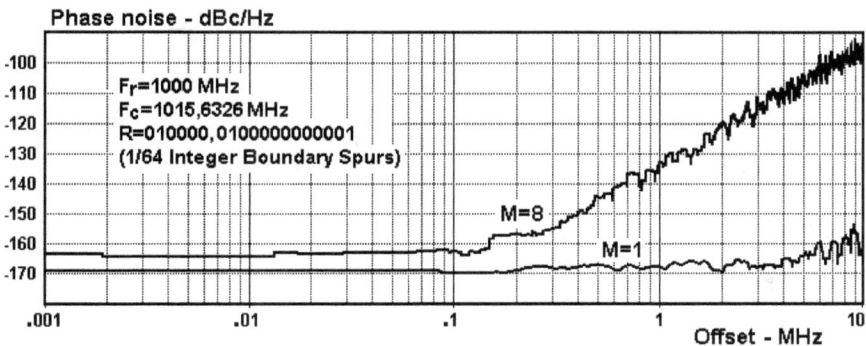

Figure 5.18 Spectrum with and without a frequency divider.

It is important to note that including the divider does not cause the appearance of multiplication in the loop of the synthesizer because the reference frequency for the main path running through pulse distributors remains the same high value.

Also there is no need for an explanation that the accumulators are nonobligatory to be of a binary capacity but may be of another format (e.g., of a decimal capacity), which it is achieving by limitation of the binary capacity. Moreover, the capacity may be changeable and controllable by programming means.

5.9 COMPARISON OF SPECTRA OF PDS, PDS-DSM, AND FRACTIONAL-N PLL SYNTHESIZERS

Figure 5.19, for comparison, shows examples of PDS, PDS-DSM. and fractional-N PLL synthesizer spectra under as many equal conditions

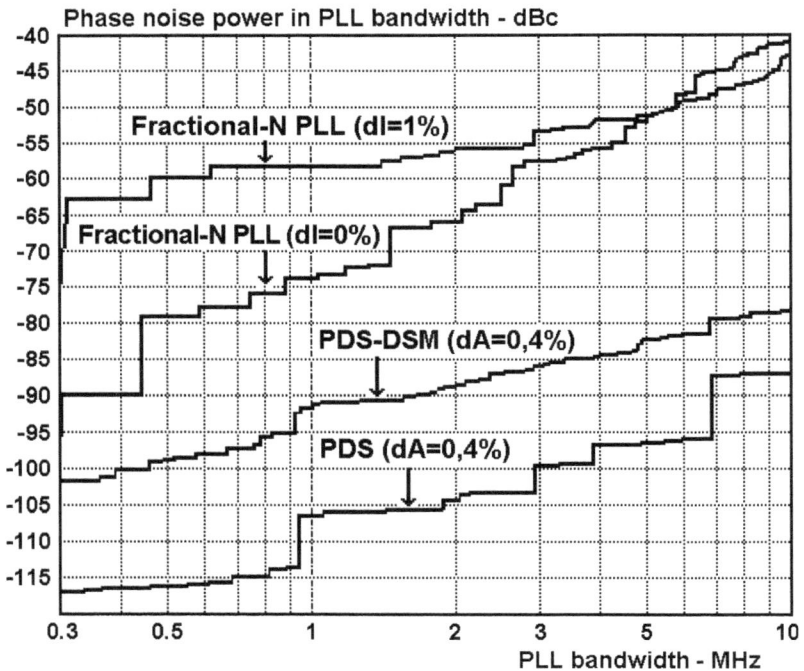

Figure 5.19 Comparison of the spectrum of different synthesizers.

as possible for one typical case where the signal frequency is $F_c = \sim2.5$ GHz.

The MFPD contains a 15-bit reference accumulator (5 bits in the MSB block and 10 bits in the LSB block). The reference frequency is $F_r = 1000$ MHz. The accumulator input control code is $R = 01000.0100000001$ (1/32 IBS). It is assumed that the inaccuracy of one of the KR bits of the DAC is 0.4%. The influence of the inaccuracy of the R2R section of the DAC on the spectrum due to its significantly lower weight is not taken into account.

There is also a diagram for the PDS-DSM variant with the same DAC inaccuracy. The block of the least significant 10 digits contains 2 accumulators for forming a DSM of the MASH-3 type. This variant is inferior to the PDS variant under such comparison conditions. However, when choosing one of the options, preference can be given to the PDS-DSM variant due to its simpler structure.

For the fractional-N PLL synthesizer, in order to obtain the same signal frequency as in the PDS, the integer division factor is chosen equal to $N_0 = 8$, and the code at the input of the least significant 10 fractional digits block is the same as in the PDS (i.e., 0100000001). The reference frequency is $F_r = 320$ MHz. Due to the presence of a frequency divider, which is mandatory in the system, it is not possible to use the 1,000-MHz reference frequency. The block of fractional digits also contains 2 accumulators for forming DSM type MASH-3. The phase detector is considered ideal in the sense of linearity of its characteristics.

In the fractional-N PLL version, a PFD with charge pump is used with an inequality of currents $dI = 1\%$ [23]. This results in additional spectrum degradation. However, it is worth noting that it is very doubtful today to obtain such a low current mismatch as 1% at such a high frequency as 320 MHz, with such large division ratio variations as 50% (with MASH-3). For example, in [17], more modest parameters are used in the simulation: comparison frequency of 12 MHz, division ratio variations within 10%, and 2% mismatch of currents.

5.10 ABOUT THE NUMBER OF SPLIT PHASES

Figure 5.20 shows, as an example for the PDS-DSM version, what happens to the fractional phase noise spectrum with an increase in the

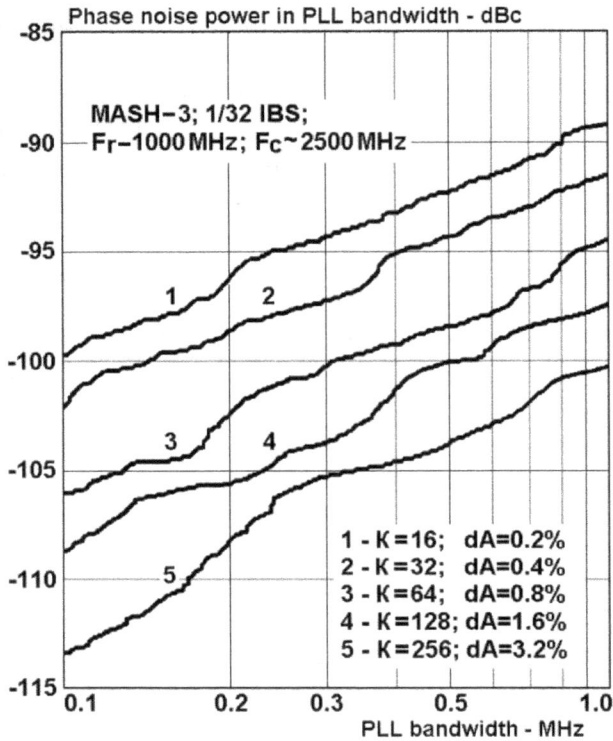

Figure 5.20 Comparison of spectrum at different values of K and dA.

number of split phases K. The spectrum is expressed as the total noise power in the PLL bandwidth, where dA in percent is the amplitude inaccuracy of one of the DAC segments. With an octave increase in K and with a simultaneous octave increase in the error dA, the spectrum not only does not deteriorate (due to a decrease in the DAC accuracy), but, on the contrary, becomes cleaner with each octave increase in the number of K split phases and a proportional decrease in the DAC accuracy.

Figure 5.21 shows the spectral density of the fractional noise for the case of $K = 128$ split phases and a DAC inaccuracy of $dA = 10\%$. The reference frequency is $F_r = 1$ GHz, and the signal frequency Fc is about 10 GHz.

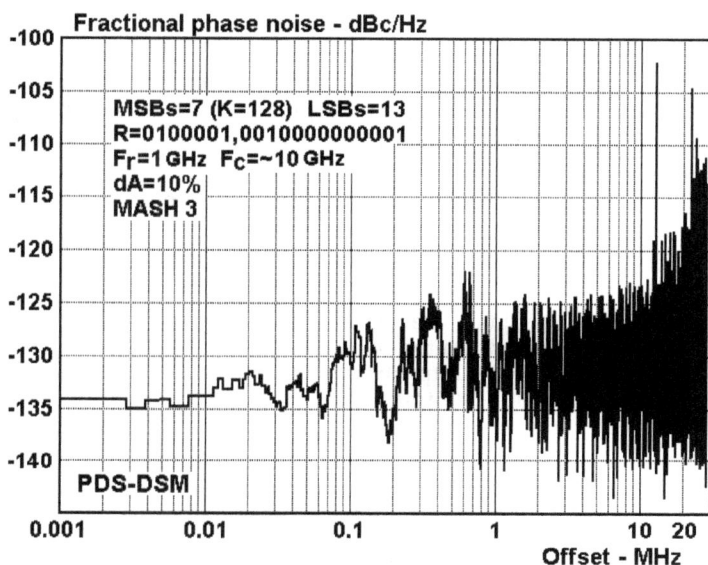

Figure 5.21 Spectral density of the fractional noise.

Extrapolation is possible. If the inaccuracy of the DAC is further increased, for example, 2 times, up to 20%, then the spectrum curve rises by 6 dB, and if it is reduced by 2 times, it will drop by the same 6 dB. The frequencies F_r and F_c can be varied similarly.

With this approach, when the high precision of the DAC is not required, an FPGA with a resistive ladder on a printed circuit board can be used. In this way, without investing significant funds, the developers can come to a simple one-loop synthesizer structure that is not inferior in parameters to complicated multiloop systems.

Note that this monograph does not mention one more possible advantage of the phase splitting method. This can provide additional noise reduction due to their incoherent summation at the outputs of the partial phase detectors. The Analog Devices Inc. website published an article [24], in which the authors experimentally, on $K = 4$ fractional-N synthesizer chips, confirmed this phenomenon, having received a gain of 6 dB in exact accordance with the formula 10lgK. Whether there will be such a gain in our case with many split phases deserves separate study.

References

[1] Rhee, W., and A. Ali, "Phase/Frequency Detector with Time-Delayed Inputs in a Charge Pump Based Phase Locked Loop and a Method for Enhancing the Phase Locked Loop Gain," US Patent 6,147,561, November 14, 2000.

[2] Chi, B., et al., *A Fractional-N PLL for Digital Clock Generation with an FIR-Embedded Frequency Divider*, Institute of Microelectronics, Tsinghua University, Beijing, China, 2007 IEEE.

[3] Tsutsumi, K., et al., *A Low Noise Multi-PFD PLL with Timing Shift Circuit*, Mitsubishi Electric Corporation, Kanagawa, Japan, 2012.

[4] Fujitsu News, "Fujitsu Develops Low-Noise Signal-Generating Circuit Technology for Automotive Radar and Other Transceivers," October 8, 2013, http://www.fujitsu.com/global/news/pr/archives/month/2013/20131008-04.html#scrollTop=0.

[5] Chang, J. -C., et al., "Phase Locked Loop with Shifted Input," US Patent 7,636,018, December 22, 2009.

[6] Koslov, V. I., "Frequency Synthesizer," Patent of Russia No. 2003227, priority 30.05.1991, published 15.11.1993, in Russian.

[7] Koslov, V. I., "Digital PLL Frequency Synthesizer," US Patent 5,748,043, May 5, 1998, PCT filed May 3, 1994, PCT/US94/04880.

[8] Varfolomeev, G. F., and V. I. Koslov, "Frequency Synthesizer for Radio Communication Equipment of the Fifth Generation," *Radio Communication Technology*, Vol. 2, 1995, in Russian.

[9] Koslov, V., "A New Concept in Frequency Synthesis," *Microwave Product Digest*, October 2010.

[10] Koslov, V., "A Low Cost PLL Frequency Synthesizer with Fine Frequency Resolution," *Microwave Product Digest*, February 2011.

[11] Chenakin, A., "Looking Beyond the Basics," *Microwave Journal*, April 2014.

[12] Koslov, V. I., "Phase-Splitting Frequency Synthesis Method in a PLL System, Part I," *Electrosvyaz*, No. 5, 2019, in Russian.

[13] Koslov, V. I., "Phase-Splitting Frequency Synthesis Method in a PLL System, Part II," *Electrosvyaz*, No. 7, 2019, in Russian.

[14] Wells J. N., "Frequency Synthesizers," US Patent 4,609,881, September 2, 1986, priority May 17, 1983, GB Patent 8,313,617.

[15] Koslov, V. I., et al., "Frequency Synthesizer with Fractional Division Ratio Modulation in a PLL," *Electrosvyaz*, No. 9, 1988, in Russian.

[16] Norsworthy, S., R. Schreier, and G. Temes, *Oversampling Delta-Sigma Data Converters*, New York: IEEE Press, 1992.

[17] Boser, B., and B. Wooley, "The Design of Sigma-Delta Modulation Analog-to-Digital Converters," *IEEE J. Solid-State Circuits*, Vol. 23, No. 6, December 1988, pp. 1298–1308.

[18] Rogers, J., C. Plett, and F. Dai, *Integrated Circuits for High-Speed Frequency Synthesis*, Norwood, MA: Artech House, 2006.

[19] Kester, W., *Analog-to-Digital Conversion*, Translation from English, Moscow: Technosphere, 2007, in Russian.

[20] King, N. J. R., "Phase-Locked Loop Variable Frequency Generator," US Patent 4,204,174, May 20, 1980, filed November 9, 1978.

[21] Miller, B. M., "Multiple Modulator-Fractional-N Divider," US Patent 5,038,117,August 6, 1991, filed September 7, 1990.

[22] Romanov, S. K., et al., "Fractional Interference in Phase-Digital Frequency Synthesizers," *Theory and Technology of Radio Communication, Scientific and Technical. Sat., Concern "Sozvezdie,"* No. 3, 2013, Voronezh.

[23] Romanov, S. K., et al., "On the Influence of the Mismatch of the Pump Currents of a Pulsed Frequency-Phase Detector on the Noise Spectrum in an PLL System with a Fractional Frequency Divider," *Theory and Technology of Radio Communication, Concern "Sozvezdie,"* Issue 1, 2008, Voronezh.

[24] Clarke, B., and J. Collins, "Replacing YIG-Tuned Oscillators with Silicon by Using an Ultrawide Band PLL/VCO with Precise Phase Control," *Analog Devices*, April 1, 2015, https://www.analog.com/en/resources/technical-articles/replacing-yig-tuned-oscillators-with-silicon.html.

6

CONCLUSION

This concludes a succinct analysis of PLL frequency synthesis methods, highlighting both their educational value and importance for advancing the field. Although PLL frequency synthesis has been covered elsewhere in existing literature, this book stands out by focusing on multifrequency phase detectors (MFPDs), a topic often only touched upon superficially in other works or entirely overlooked. The investigation of MFPDs for phase comparison across multiple frequencies introduces new, highly technical content that deepens our understanding of these essential components in modern communication systems.

MFPDs are crucial for various complex tasks, such as clock and data recovery in high-speed communication systems, frequency synthesis in RF transceivers, and PLL operation across different electronic systems. Their ability to effectively handle multiple frequencies is vital in the modern field of communication and signal processing. A key advantage of MFPDs is their design, which removes the division ratio in the PLL, resulting in significant reductions in noise and unwanted spurious signals. This design innovation helps to achieve exceptional spectral purity, high speed, superior resolution, and low-power consumption, all while keeping costs down. The text showcases a forward-looking approach to frequency synthesis, offering substantial potential for future applications in the field.

The reader has now been provided with a comprehensive overview of PLL frequency synthesis methods, with a special focus on

MFPDs and valuable insights for practitioners and researchers working on high-frequency synthesizers and clock generators. Unlike more general reference texts, this work concentrates on offering practical guidance for those striving to deepen their understanding of advanced PLL techniques in real-world applications.

SELECTED BIBLIOGRAPHY

Chang, J. -C., et al., "Phase Locked Loop and Method Thereof," US Patent 2008/0224789, Assignee–United Microelectronics Corp., September 18, 2008, filed March 14, 2007.

Chang, J. -C., et al., "Phase Locked Loop with Phase Shifted Input," US Patent 7636018, Assignee–United Microelectronics Corp., December 22, 2009, filed March 14, 2007.

Gillette, G. C., "Frequency Synthesizer System," US Patent 3,582,810, June 1, 1971, filed May 5, 1969.

Kim, K. -J., et al., "Phase-Locked Loop Based Frequency Synthesizer and Method of Operating the Same," US Patent 8,373,469 B2, Assignee–Korea Electronics Technology Institute, February 12, 2013filed December 30, 2010.

Kim, Y. H., et al., "Fractional-N Frequency Synthesizer and Method Thereof," US Patent 2010/0321120 A1, Assignee–Samsung Electro-Mechanics Co., Ltd., December 23, 2010, filed September 21, 2009.

Rhee, W., et al., "Frequency Divider, Frequency Synthesizer and Application Circuit," US Patent 2010/0225361, Assignee–Samsung Electronics Co., Ltd., September 9, 2010, filed July 9, 2009.

Soo, K. S., "Frequency Synthesizer and Polar Transmitter Having the Same," US Patent 2010/0329388, Assignee–Ko Sang Soo, 2010, December 30, 2010, filed April 16, 2010.

Tajalli, A., "High Performance Phase Locked Loop," US Patent 10,057,049 B2, Assignee–Kandou Labs, S.A., August 21, 2018, filed April 21, 2017.

Unruch, G. A., "Apparatus and Method for Combining Multiple Charge Pumps in Phase Locked Loops," US Patent 9,520,889 B2, Assignee–Avago Technologies International Sale, December 13, 2016, filed February 19, 2015.

ABOUT THE AUTHOR

Vitaly Koslov is an accomplished independent researcher and expert in radio frequency and microwave engineering. He earned his degree in radio equipment development from the Taganrog Radio Engineering Institute in 1959 and later defended his Candidate of Technical Sciences dissertation at the Tomsk Institute of Radio Physics in 1968. Over the course of his distinguished career, Dr. Koslov has specialized in frequency synthesis and phase-locked loop (PLL) technologies, with a particular focus on multifrequency phase detectors (MFPDs).

Dr. Koslov has held research positions at leading institutes in Omsk and Kiev, contributing significantly to the advancement of PLL-based frequency synthesizers. His work is recognized internationally, with publications in both Russian and American scientific journals, and he holds several patents across both countries. He continues to influence the field from his home in Kiev, Ukraine, where he remains an active contributor to the global engineering community.

INDEX

Artech House Microwave Library

System-in-Package RF Design and Applications, Michael P. Gaynor

Technologies for RF Systems, Terry Edwards

Terahertz Metrology, Mira Naftaly, editor

Understanding Quartz Crystals and Oscillators, Ramón M. Cerda

Vertical GaN and SiC Power Devices, Kazuhiro Mochizuki

The VNA Applications Handbook, Gregory Bonaguide and Neil Jarvis

Wideband Microwave Materials Characterization, John W. Schultz

Wired and Wireless Seamless Access Systems for Public Infrastructure, Tetsuya Kawanishi

For further information on these and other Artech House titles, including previously considered out-of-print books now available through our In-Print-Forever® (IPF®) program, contact:

Artech House
685 Canton Street
Norwood, MA 02062
Phone: 781-769-9750
Fax: 781-769-6334
e-mail: artech@artechhouse.com

Artech House
16 Sussex Street
London SW1V 4RW UK
Phone: +44 (0)20 7596 8750
Fax: +44 (0)20 7630 0166
e-mail: artech-uk@artechhouse.com

Find us on the World Wide Web at: www.artechhouse.com

www.ingramcontent.com/pod-product-compliance
Lightning Source LLC
Chambersburg PA
CBHW050502190326
41458CB00005B/1396